T0260979

Formal Verification of Control
System Software

PRINCETON SERIES IN APPLIED MATHEMATICS

Ingrid Daubechies (Duke University); Weinan E (Princeton University); Jan Karel Lenstra (Centrum Wiskunde & Informatica, Amsterdam); Endre Süli (University of Oxford)

The Princeton Series in Applied Mathematics publishes high quality advanced texts and monographs in all areas of applied mathematics. Books include those of a theoretical and general nature as well as those dealing with the mathematics of specific applications areas and real-world situations.

A list of titles in this series appears at the back of the book.

Formal Verification of Control System Software

Pierre-Loïc Garoche

PRINCETON UNIVERSITY PRESS

PRINCETON AND OXFORD

Published by Princeton University Press
41 William Street, Princeton, New Jersey 08540
6 Oxford Street, Woodstock, Oxfordshire OX20 1TR

press.princeton.edu

LCCN 2019930479

ISBN 978-0-691-18130-1

British Library Cataloging-in-Publication Data is available

Editorial: Vickie Kearn, Susannah Shoemaker, and Lauren Bucca
Production Editorial: Brigitte Pelner
Jacket/Cover credit: Sculpture by Patrick Meichel
Production: Erin Suydam
Publicity: Alyssa Sanford

This book has been composed in LATEX

Printed on acid-free paper ∞

Printed in the United States of America

10 9 8 7 6 5 4 3 2 1

To Pierre Garoche: the captain, the engineer, and the researcher

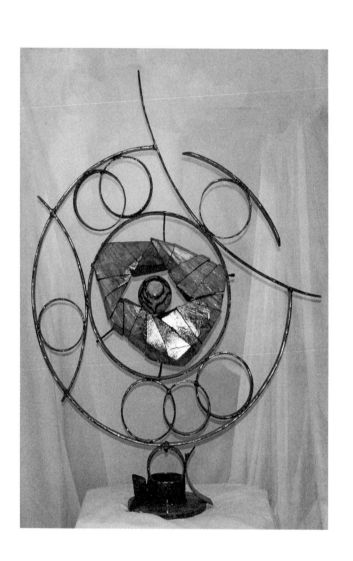

Contents

Part I

Need and Tools to Verify Critical Cyber-Physical Systems

Chapter One

Critical Embedded Software
Control Software Development and V&V

CYBER-PHYSICAL SYSTEMS (CPS) is a kind of buzzword capturing the set of physical devices controlled by an onboard computer, an embedded system. Critical embedded systems are a subset of these for which failure is not acceptable. Typically this covers transportation systems such as cars, aircraft, railway systems, space systems, or even medical devices, all of them either for the expected harmfulness for people, or for the huge cost associated with their failure.

A large part of these systems are controllers. They are built as a large running loop which reads sensor values, computes a feedback, and applies it to the controlled system through actuators. For most systems, at least in the aerospace industry, the time schedule for controllers is so tight that these systems have to be "real time." The way these systems have been designed requires the execution of the loop body to be performed within some time to maintain the system in a reasonable state. In the civil aircraft industry, the controller itself is rather complex, but is built as a composition of simpler controllers. Furthermore, the global system accounts for potential failures of components: sensors, networks, computers, actuators, etc., and adapts the control to these discrepancies.

The increase of computer use in those systems has led to huge benefits but also an exponential growth in complexity. Computer based systems compared to analog circuits enable more efficient behaviors, as well as size and weight reductions. For example, aircraft manufacturers are building control laws for their aircraft that maintain them at the limit of instability, allowing more fuel efficient behavior;[1] Rockwell Collins implemented a controller for a fighter aircraft able to recover controllability when the aircraft loses, in flight, from 60 to 80% of one of its wings;[2] United Technology has been able to replace huge and heavy

---

[1] In an A380, fuel is transferred between tanks to move the center of gravity to the aft (backward). This degrades natural stability but reduces the need for lift surfaces and therefore improves fuel efficiency by minimizing total weight and drag. See the book *Airbus A380: Superjumbo of the 21st Century* by Noris and Wagner [58].

[2] Search for Damage Tolerance Flight Test video, e.g., at https://www.youtube.com/watch?v=PTMpq_8SSCI

power electric systems with their electronic counterpart, with a huge reduction in size and weight.[3]

The drawback of this massive introduction of computers to control systems is the lack of predictability for both computer and software. While the industry has been accustomed to having access to the precise characteristic of its components, e.g., a failure rate for a physical device running in some specific conditions, these figures are hardly computable for software, because of the intrinsic complexity of computer programs.

Still, all of us are now used to accepting software licenses where the software vendor assumes nothing related to the use of the software and its possible impact. These kinds of licenses would be unacceptable for any other industry.

To conclude with this brief motivation, the aerospace industry, and more generally critical embedded systems industries, is are now facing a huge increase in the software size in their systems. This is motivated first by system complexity increases because of safety or performance objectives, but also by the need to integrate even more advanced algorithms to sustain autonomy and energy efficiency. **Guaranteeing the good behavior of those systems is essential to enable their use.**

Until now, classical means to guarantee good behavior were mainly relying on tests. In the aerospace industry the development process is strictly constrained by norms such as the DO-178C [104] specifying how to design software and perform its verification and validation (V&V). This document shapes the V&V activities and requires the verification to be specification-driven. For each requirement expressed in the design phases, a set of tests has to be produced to argue that the requirement is satisfied. However, because of the increase in complexity of the current and future systems, these test-based verifications are reaching their limit. As a result the cost of V&V for systems has exploded and the later a bug is found, the more expensive it is to solve.[4]

Last, these certification documents such as DO-178C have been recently updated, accounting for the recent applicability of formal methods to argue about the verification of a requirement. Despite their possible lack of results in a general setting, these techniques, in cases of success, provide an exhaustive result, i.e., they guarantee that the property considered is valid for all uses, including systems admitting infinite behaviors.

All the works presented in this book are motivated by this context. We present formal methods sustaining the verification of controller properties at multiple stages of their development. The goal is to define new means of verification, specific to controller analysis.

---

[3]E.g., Active EMI filtering for inverters used at Pratt and Whitney, Patent US20140043871.

[4]USA NIST released in 2002 an interesting survey, "The Economic Impacts of Inadequate Infrastructure for Software Testing," detailing the various costs of verification and bugs. Chapter 6 is focused on the transportation industry.

CURRENT LIMITS & OBJECTIVES

The objectives of the presented works are restricted to the definition of formal methods-based analyses to support the verification of controller programs.

More specifically we can identify the following limits in the current state of the art:

Need to compute invariants of dynamical systems New advances in formal methods are often not specialized for a particular kind of program. They rather try to handle a large set of programming language constructs and deal with scalability issues. In specific cases, such as the application of static analysis to Airbus programs [54], dedicated analyses, like the second-order filter abstraction [47], have been defined. But the definition of these domains is tailored to the program for which they are defined.

Lack of means to compute nonlinear invariants As we will see in this book, the simplest properties of controllers are often based on at least quadratic properties. Again, because of efficiency and scalability, most analyses are bound to linear properties. We claim that more expressive yet more costly analyses are required in specific settings such as the analysis of control software. The scalability issues have to be addressed by carefully identifying the local part of the program on which to apply these more costly analyses.

Expressivity of static analysis properties Formal methods applied at model or code level are hardly used to express or analyze system-level properties. In practice, static analysis is mainly bound to numerical invariants while deductive methods or model-checking can manipulate more expressive first-order logic formulas. However, computer scientists are usually not aware of the system-level properties satisfied or to be satisfied by the control program they are analyzing. An important research topic is therefore the use of these formalisms (first-order logic and numerical invariants) to express and analyze system-level properties.

Scope of current analyses In the current state of the practice, concerns are split and analyzed locally. For example the control-level properties such as stability are usually analyzed by linearizing the plant and the controller description. At the code level this can be compared to the analysis of a simplified program without if-then-else or nonlinear computations. Similarly, the complete fault-tolerant architecture, which is part of the implemented embedded program, is abstracted away when analyzing system-level properties. A last example of such—potentially unsound—simplifications, is the assumption of a real semantics when performing analyses, while the actual implementation will be executed with floating-point semantics and the associated errors. The vision supported by the book is that more integrated analyses should address the study of the global system.

The proposal is mainly developed in two complementary directions:

- nonlinear invariant synthesis mainly based on the use of convex optimization techniques;
- consideration of system-level properties on discrete representation, at code level, with floating-point semantics.

This book is structured in four parts:

Part I introduces formal methods and controller design. It is intended to be readable both by a control scientist unaware of formal methods, and by a computer scientist unaware of controller design. References are provided for more scholastic presentations.

Part II focuses on invariant synthesis for discrete dynamical systems, assuming a real semantics. All techniques are based on the computation of an inductive invariant as the resolution of a convex optimization problem.

Part III revisits basic control-level properties as numerical invariants. These properties are typically expressed on the so-called *closed-loop representation*. In these chapters we assume that the system description is provided as a discrete dynamical system, without considering its continuous representation with ordinary differential equations (ODEs).

Part IV extends the previous contributions by considering floating-point computations. A first part considers that the program analyzed is executed with floating-point semantics and searches for an inductive invariant considering the numerical errors produced. A second part ensures that the use of convex optimization, a numerical technique, does not suffer from similar floating-point errors.

Chapter Two

Formal Methods

Different Approaches for Verification

WHILE TESTING IS A COMMON practice for a lot of engineers as a way to evaluate whether the program they developed fulfills its needs, formal methods are less known and may require a little introduction to the non-expert. This chapter can be easily skipped by the formal verification reader but should be a reasonable introduction to the control expert engineer.

In this chapter we give a brief overview of some of these formal methods and their use in the context of critical embedded systems development. We first define the semantics of programs: their basic properties and their meaning. Then, we outline different formal verifications and explain how they reason on the program artifact. A last part addresses the soundness of the analyses with respect to the actual semantics.

2.1 SEMANTICS AND PROPERTIES

Let us first consider a simple imperative program as we could write in C code and use it to introduce basic notions:

```c
1  int f (x) {
2      int y = 42;
3      while (x > 0) {
4          x = x - 2;
5          y = y + 4;
6      }
7      return y;
8  }
```

For a given input x, this program is deterministic: it admits a single execution. Let us assume it is called with $x = 3$. In that case the execution is finite

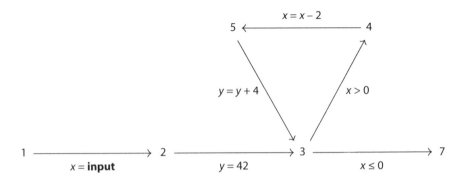

Figure 2.1. Control flow graph.

and will stop once x becomes nonpositive, here $x = -1$. This happens after two executions of the loop body. Therefore, $y = 42 + (2 * 4) = 50$ when the program stops.

The semantics, i.e., the meaning of this program, can be characterized in different ways. One approach is to see it as a function that takes inputs—here x—and returns the output value y. We speak of a *denotational semantics*: $[\![f]\!]_{\text{den}}(3) = 50$. We could characterize the output of the program f as a mathematical function $f : \mathbb{Z} \to \mathbb{Z}$ of x:

$$f(x) = 42 + 4 * \left\lceil \frac{x}{2} \right\rceil.$$

Another approach details the steps of the computation and does not only focus on the result. This is *operational semantics*. In operational semantics, one describes the behavior of the program as a sequence of transitions between states. A state denotes a current view of the program. In this simple case, a state can be characterized by a triple program point (pp), x, and y. In the following, we denote by Σ such a set of states. Let us look at the simple execution of f with input 5:

state	0	1	2	3	4	5	6	7	8
pp	2	3	4	5	3	4	5	3	7
x	5	5	3	3	3	1	1	1	-1
y	42	42	42	46	46	46	50	50	50

The run of the program is here described by a sequence of states, a trace: $s_0 \to s_1 \to \ldots \to s_8$. In this case of a deterministic function, each trace is only characterized by its initial element s_0. Initial elements are a subset of states: let $Init \subseteq \Sigma$ be such a set. The set of rules describing possible transitions from one state to the other characterizes the operational semantics of the program. Let us denote it by $[\![f]\!]_{\text{op}} \in \Sigma \times \Sigma$, the set of transitions from state to state. One can also represent it as a kind of automaton: the control flow graph (see Figure 2.1).

By interpreting a program as a set of states Σ, an initial set of states $Init \subseteq \Sigma$, and a transition relation $[\![\cdot]\!]_{op} \subseteq \Sigma \times \Sigma$, we defined a transition system $(\Sigma, Init, [\![\cdot]\!]_{op})$.

In practice, one is not necessarily interested directly in the program semantics in a denotational or operational form but rather by the properties of the program when executed.

The most precise definition of a program behavior is to characterize exactly its set of traces, its *trace semantics*:

$$[\![f]\!]_{trace} = \left\{ s_0 \to \ldots \to s_n \;\middle|\; \begin{array}{l} \forall i \in [0, n-1], (s_i, s_{i+1}) \in [\![f]\!]_{op} \\ s_0 \in Init \end{array} \right\}.$$

In the case of nonterminating programs, traces could be infinite. While non-terminating programs are usually seen as bad programs in computer science, controllers are supposed to be executed without time limit, in a while true loop. We can extend the definition of trace semantics for infinite traces:

$$[\![f]\!]_{trace} = \left\{ s_0 \to \ldots \to s_i \to \ldots \;\middle|\; \begin{array}{l} \forall i \geq 0, (s_i, s_{i+1}) \in [\![f]\!]_{op} \\ s_0 \in Init \end{array} \right\}.$$

To summarize, trace semantics capture the possibly infinite set of possibly infinite traces. If provided with such a set, one can observe any properties related to intermediate computed values, occurrence of states within traces, infinite behavior, finite behavior such as deadlocks, and so on. These properties are usually defined as temporal properties.

Another semantics of interest, with respect to the program semantics, is the *collecting semantics*. This semantics focuses only on reachable states in traces but not on their specific sequences.

One can define it as follows:

$$[\![f]\!]_{coll} = \left\{ s_n \;\middle|\; \begin{array}{l} \exists\, s_0 \to \ldots \to s_n \in [\![f]\!]_{trace} \text{ i.e. such that} \\ \exists s_0, \ldots, s_n, \ldots \in \Sigma \\ \forall i \in [0, n-1], (s_i, s_{i+1}) \in [\![f]\!]_{op} \\ s_0 \in Init \end{array} \right\}.$$

As such, collecting semantics is an abstraction of trace semantics: it characterizes a set of reachable states but loses information on their relationship. This semantics is, however, extremely useful: it can capture all reachable states and therefore guarantee that all such states verify a given invariant, or avoid a given bad region.

A last way to express the behavior of a program is the axiomatic semantics. This notion was introduced by Hoare in 1969 [4]. In 1967 Floyd [3] proposed to annotate a flowchart by its local invariants. The following figure is extracted from that paper.

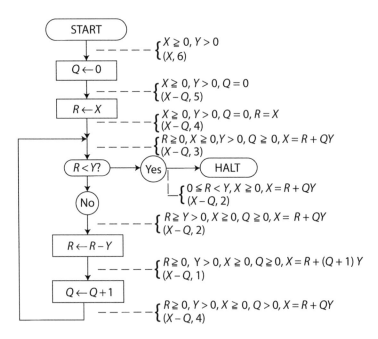

$$\begin{cases} X \geq 0, Y > 0 \\ (X, 6) \end{cases}$$

$$\begin{cases} X \geq 0, Y > 0, Q = 0 \\ (X - Q, 5) \end{cases}$$

$$\begin{cases} X \geq 0, Y > 0, Q = 0, R = X \\ (X - Q, 4) \end{cases}$$

$$\begin{cases} R \geq 0, X \geq 0, Y > 0, Q \geq 0, X = R + QY \\ (X - Q, 3) \end{cases}$$

$$\begin{cases} 0 \leq R < Y, X \geq 0, X = R + QY \\ (X - Q, 2) \end{cases}$$

$$\begin{cases} R \geq Y > 0, X \geq 0, Q \geq 0, X = R + QY \\ (X - Q, 2) \end{cases}$$

$$\begin{cases} R \geq 0, Y > 0, X \geq 0, Q \geq 0, X = R + (Q + 1)Y \\ (X - Q, 1) \end{cases}$$

$$\begin{cases} R \geq 0, Y > 0, X \geq 0, Q > 0, X = R + QY \\ (X - Q, 4) \end{cases}$$

Figure 2.2. Assigning meanings to programs by Floyd.

In [4], "An Axiomatic Basis for Computer Programming," Hoare defines a deductive reasoning to validate code level annotations. This paper introduces the concept of Hoare triple $\{Pre\}code\{Post\}$ as a way to express the semantics of a piece of code by specifying the postconditions ($Post$) that are guaranteed after the execution of the code, assuming that a set of preconditions (Pre) was satisfied. Hoare supports a vision in which this axiomatic semantics is used as the "ultimately definitive specification of the meaning of the language..., leaving certain aspects undefined.... Axioms enable the language designer to express its general *intentions* quite simply and directly, without the mass of detail which usually accompanies algorithmic descriptions."

Assuming the Euclidian division algorithm presented in Fig. 2.2 is implemented in a C function $\text{div}(x,y,*q,*r)$, one can specify the contract as follows:

```
void div(x,y,*q,*r) {
    // { x≥0 ∧ y>0 }
    *q = 0;
    *r = x;
    while (*r < y) { ... };
    // { 0 ≤*r<y ∧ x≥0 ∧ x=*r+*q×y }
}
```

As envisioned by Hoare, this approach has been largely developed and is used to specify formally the intended behavior of a program as a set of Hoare triples. Theoretically speaking, axiomatic semantics is a further abstraction of operational or denotational semantics since it only constrains valid implementations.

2.2 A FORMAL VERIFICATION METHODS OVERVIEW

We now illustrate the basic principles behind main verification methods: deductive methods (DM), SMT-based model-checking (MC), and abstract interpretation (AI). First, we sketch here how these techniques work on simple loopless examples. Then, we elaborate more on some details of their implementation or their use on more realistic examples. The exhaustive presentation of these techniques, developed in the last thirty to forty years, cannot be done in a few pages. The presentation reflects the author's view and understanding of these approaches.

First, let us make a disappointing statement:

Theorem 2.1 (Rice's theorem). *It is undecidable to determine whether the language recognized by an arbitrary Turing machine T lies in a nontrivial set of languages S.*

$$\mathcal{L}(T) \subseteq S \text{ is undecidable}$$

where $\mathcal{L}(T)$ denotes the language recognized by the Turing machine T.

This theorem, which may not be easily readable for the theoretical computer science agnostic, states that any property of interest is hardly analyzable on a program. In other words: "It is undecidable to determine whether an arbitrary program satisfies a nontrivial property."

Because of undecidability it is worthless to design sound, complete, and terminating techniques for arbitrary programs and properties. Let us denote by $Prog \models P$ the validity of property P for program $Prog$ and by $Prog \vdash_A P$ the fact that the analysis A stated that P was valid for program $Prog$. We can define, for all program $Prog$ and property P:

$$Prog \vdash_A P \Rightarrow Prog \models P \qquad \text{(soundness)}$$
$$Prog \models P \Rightarrow Prog \vdash_A P \qquad \text{(completeness)}$$
$$Prog \vdash_A P \text{ terminates.}$$

Formal verification techniques usually address this issue by focusing on sound and terminating methods, that is, without the completeness property. This amounts to computing an intermediate stronger property P' such that

$$((Prog \vdash_A P') \wedge (P' \Rightarrow P)) \Rightarrow Prog \models P$$
$$Prog \vdash_A P' \text{ terminates.}$$

This is often referred to as over-approximation techniques, or conservative techniques: showing the validity of P amounts to computing a less precise property P' which may imply P. Even if the property P was actually valid on the program, the lack of precision of P' may not permit us to prove $P' \implies P$, leading to a lack of conclusion: P has an unknown status for program $Prog$, and the analysis has been unable to conclude with respect to P.

Remark 2.2 (Termination of analysis vs. program). *Note that termination of analysis is unrelated to the existence of infinite traces in the analyzed program. A nonterminating formal verification technique may fail to return a result on a finite transition system admitting only finite traces, while a terminating analysis will conclude even for systems admitting infinite behaviors.*

2.2.1 Basic principles illustrated on a loopless example

Let us first focus on a simple loopless example, for example, the infinity norm in $\mathbb{R}^2$:

```
 1   real norminf (real x, y) {
 2     real xm, ym, r;
 3     if (x >= 0) // compute abs(x)
 4       {xm = x;}
 5     else
 6       {xm = -x;};
 7     if (y >= 0) // compute abs(y)
 8       {ym = y;}
 9     else
10       {ym = -y;};
11     if (xm >= ym) // compute max(xm, ym)
12       {r = xm;}
13     else
14       {r = ym;};
15     return r;
16   }
```

We are interested in the following properties:

- null on zero: $norminf(0,0) = 0$;
- positivity: $\forall(x,y), norminf(x,y) \geq 0$.

Note that, in that case, the formalization of the specification, that is, the properties of interest, as formal artifacts was rather straightforward. It may be more difficult when considering natural language description with ambiguous statements. This is another added value of formal methods: disambiguation of specification by imposing the need of strict formalization.

Of course, a first classical approach could rely on tests to evaluate the validity of these properties. We will see how various formal method techniques reason on that program, trying to prove the desired properties:

- DM: use of predicate transformation, either forward or backward reasoning;
- MC: propositional encoding and SMT-based reasoning;
- AI: interpretation of each computation in an abstract domain.

2.2.1.1 Deductive methods: predicate transformers

Deductive methods are the evolution of the ideas proposed by Hoare [4]. Predicate transformation allows us to apply the semantics of the considered program on the formal representation of the property. These manipulations can be performed either in a forward manner, transforming the precondition through the code—we speak about *strongest postcondition*—or, in the opposite direction, propagating back the postcondition through the code—we speak about *weakest precondition*. While both techniques should be equally sound, most implementations used in C code analysis [43, 67, 98, 108] rely on the weakest precondition algorithm.

This method computes $wp(code, Post)$, the weakest precondition such that, when executing the code, $Post$ is guaranteed. The rules are defined on the structure on the imperative code, per statement kind and applied iteratively. On naive imperative languages statements can be either assignments or control structures such as sequencing of statements, conditionals (if-then-else), or loops.

The assignment rule amounts to substituting in the postcondition B any occurrence of x by its definition e:

$$wp(x := e, B) \triangleq [e/x]B.$$

Example 2.3. *Let us illustrate this mechanism on the simplest example. Assume the postcondition requires $y \leq 0$. The weakest precondition of the instruction $y = x+1$; imposes $x \leq -1$.*

```
// { x+1 ≤ 0 ≡ x ≤ -1 }                    (C)
y = x + 1;
// { y ≤ 0 }
```

Weakest precondition composes well: once the computation of the impact of c_2 to B has been computed, it can be used to propagate the impact of statement c_1:

$$wp(c_1; c_2, B) \triangleq wp(c_1, wp(c_2, B)).$$

Conditional statements (if-then-else) are encoded as a disjunction: one obtains B after executing the statement, either because b holds and c_1 gives B, or because $\neg b$ holds and c_2 gives B:

$$wp(\text{if } b \text{ then } c_1 \text{ else } c_2, B) \triangleq \wedge \quad \begin{aligned} & b \Rightarrow wp(c_1, B), \\ & \neg b \Rightarrow wp(c_2, B). \end{aligned}$$

In our example, we have two properties expressed as the following Hoare triples:

$$\{(x, y) = (0, 0)\} \ norminf \ \{\backslash r = 0\}, \tag{2.1}$$

$$\{True\} \ norminf \ \{\backslash r \geq 0\}. \tag{2.2}$$

The first Hoare triple states that when $(x, y) = (0, 0)$ the result is 0, while the second one makes no assumption on the input: it should be valid in any context.

Let us look, manually, at this computation on the first property:

$$\backslash r = 0$$

is transformed through the last statement, a conditional statement (ite) on line 11. We obtain the weakest precondition of the statement line 11 guaranteeing $\backslash result = 0$. Each then and else block is analyzed with the wp algorithm, producing the required predicate $xm = 0$ or $ym = 0$. Then the weakest precondition of the conditional statement is produced:

$$(xm \geq ym \Rightarrow xm = 0) \wedge (xm < ym \Rightarrow ym = 0).$$

This predicate is further transformed in the leaves of the previous statement, at line 7. Then, block at line 8 is associated to the weakest precondition:

$$(xm \geq y \Rightarrow xm = 0) \wedge (xm < y \Rightarrow y = 0),$$

while the else-block at line 10 gives

$$(xm \geq -y \Rightarrow xm = 0) \wedge (xm < -y \Rightarrow -y = 0).$$

Combined with the conditional rule, this gives:

$$\left(y \geq 0 \Rightarrow \left(\begin{aligned} & (xm \geq y \Rightarrow xm = 0) \\ & \wedge (xm < y \Rightarrow y = 0) \end{aligned} \right) \right)$$
$$\wedge \left(y < 0 \Rightarrow \left(\begin{aligned} & (xm \geq -y \Rightarrow xm = 0) \\ & \wedge (xm < -y \Rightarrow -y = 0) \end{aligned} \right) \right).$$

Let us again propagate this weakest precondition to the previous statement at line 3. We obtain for its then-block the predicate

$$\left(y \geq 0 \Rightarrow \left(\begin{array}{c} (x \geq y \Rightarrow x = 0) \\ \wedge (x < y \Rightarrow y = 0) \end{array} \right) \right)$$
$$\wedge \left(y < 0 \Rightarrow \left(\begin{array}{c} (x \geq -y \Rightarrow x = 0) \\ \wedge (x < -y \Rightarrow -y = 0) \end{array} \right) \right)$$

and for its else-block

$$\left(y \geq 0 \Rightarrow \left(\begin{array}{c} (-x \geq y \Rightarrow -x = 0) \\ \wedge (-x < y \Rightarrow y = 0) \end{array} \right) \right)$$
$$\wedge \left(y < 0 \Rightarrow \left(\begin{array}{c} (-x \geq -y \Rightarrow -x = 0) \\ \wedge (-x < -y \Rightarrow -y = 0) \end{array} \right) \right).$$

Last, the conditional rule is applied:

$$\left(\left(x \geq 0 \Rightarrow \left(\left(y \geq 0 \Rightarrow \left(\begin{array}{c} (x \geq y \Rightarrow x = 0) \\ \wedge (x < y \Rightarrow y = 0) \end{array} \right) \right) \wedge \left(y < 0 \Rightarrow \left(\begin{array}{c} (x \geq -y \Rightarrow x = 0) \\ \wedge (x < -y \Rightarrow -y = 0) \end{array} \right) \right) \right) \right) \right. \\ \wedge \left(x < 0 \Rightarrow \left(\left(y \geq 0 \Rightarrow \left(\begin{array}{c} (-x \geq y \Rightarrow -x = 0) \\ \wedge (-x < y \Rightarrow y = 0) \end{array} \right) \right) \wedge \left(y < 0 \Rightarrow \left(\begin{array}{c} (-x \geq -y \Rightarrow -x = 0) \\ \wedge (-x < -y \Rightarrow -y = 0) \end{array} \right) \right) \right) \right) \right). \tag{2.3}$$

This large predicate represents the weakest precondition, that, when satisfied, is guaranteed to obtain $\backslash r = 0$ after executing the code. In this first property, the precondition was $(x, y) = (0, 0)$. Therefore, we have to prove

$$(x, y) = (0, 0) \Rightarrow (2.3).$$

This proof is sent to a satisfiability modulo theory solver (SMT) such as Alt-Ergo [78], Z3 [82], CVC4 [65, 132], or Yices [60, 133]. These solvers extend a SAT core to predicates whose atoms are expressed in other (numerical) theories.

In this specific case, the formula is easily analyzed-it can even be done by hand-and reduces to the predicate

$$True.$$

The second property can be similarly analyzed and will generate the following proof objective

$$True \Rightarrow \{(2.3) \text{ in which } v = 0 \text{ becomes } v \geq 0\}.$$

2.2.1.2 SMT-based model-checking: propositional encoding and satisfiability

SMT-based model-checking will perform similarly on this specific example. The idea is to map all constructs as predicates. One can, for example, rename variables to avoid multiple assignments to the same variable. This amounts to embeding the imperative program as functional dependencies between input and output. Let $[\![norminf]\!]_{MC}(x, y, r)$ be such function.

The proof objectives become:

$$(x = 0 \wedge y = 0) \wedge [\![norminf]\!]_{MC}(x, y, r) \wedge (r = 0) \qquad (2.4)$$
$$[\![norminf]\!]_{MC}(x, y, r) \wedge (r \geq 0). \qquad (2.5)$$

The difference with deductive method is not really sensible on this oversimple example. The main one is that no order is specified on the model-checking approach, while weakest precondition rules do transform the predicate statement after statement. One can also notice that the expression of the functional representation of $[\![norminf]\!]_{MC}(x, y, r)$ is identical in both properties (2.4) and (2.5). In deductive method, the form of the predicate representation of the code widely depends on the property analyzed.

In both cases, the final validity of the propositional encoding of the property is delegated to external solvers such as SMT solvers.

2.2.1.3 Abstract interpretation (of collecting semantics): over-approximating reachable states

Abstract interpretation relies on a different algorithm. We will develop it in its general setting in the next section. In contrast to previous methods which are able to represent complex properties through logical predicates but rely on external solvers to determine the satisfiability of these formulas, abstract interpretation paradigm intends to restrict a priori the form of the properties manipulated, providing constructive means to analyze them.

These constrained properties are called abstract domains and, since we are focused on the abstraction of the collecting semantics, they represent set of states. One can see an abstract domain D as a subset of set of states: $D \subseteq \wp(\Sigma)$.

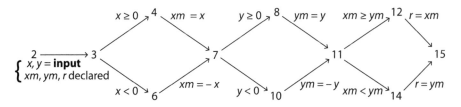

Figure 2.3. Control flow graph for infinity norm.

A classical example—and a widely used one—is the abstract domain of intervals $\mathcal{M}$ to represent subsets of $\mathbb{R}$ and the use of interval arithmetic to manipulate these abstract values. An abstract environment is used to represent a set of states. Let us informally show the computation of the abstract environment in our example before providing more theoretical background.

The computations are performed on the control flow graph. Figure 2.3 characterizes it for our example.

One can associate to each program point its abstract collecting semantics equations. These equations define the local abstract environment, depending on the predecessor values:

$$S_2 = \{\text{any value}\}$$

$$S_3 = S_2 \begin{bmatrix} x & \mapsto &]-\infty, +\infty[\\ y & \mapsto &]-\infty, +\infty[\\ xm & \mapsto &]-\infty, +\infty[\\ ym & \mapsto &]-\infty, +\infty[\\ r & \mapsto &]-\infty, +\infty[\end{bmatrix}$$

$$S_4 = S_3[x \geq 0]$$
$$S_6 = S_3[x < 0]$$
$$S_7 = S_4 \sqcup S_6$$
$$S_8 = S_7[y \geq 0]$$
$$S_{10} = S_7[y < 0]$$
$$S_{11} = S_8 \sqcup S_{10}$$
$$S_{12} = S_{11}[xm \geq ym]$$
$$S_{14} = S_{11}[xm < ym]$$
$$S_{15} = S_{12} \sqcup S_{14}$$

where $S[e > 0]$ denotes the environment S in which the abstract evaluation of e is constrained to be positive; $S[x \mapsto e]$ denotes the environment S in which variable x is updated to the abstract value e; and $S_1 \sqcup S_2$ denotes the lift of interval join to maps: $[x \mapsto S_1(x) \cup_{\mathcal{M}} S_2(x)]$.

When entering in the function body, nothing is assumed on x and y. The abstract environment is then the map

$$x \mapsto \ ]-\infty, +\infty[\qquad y \mapsto \ ]-\infty, +\infty[.$$

Depending on the language semantics, the declaration of local variables at line 2 can either assign them to a default value, or, like in C, give a value. We have the following updated abstract environment at line 3:

$$
\begin{array}{llll}
x & \mapsto &]-\infty, +\infty[& y & \mapsto &]-\infty, +\infty[\\
xm & \mapsto &]-\infty, +\infty[& ym & \mapsto &]-\infty, +\infty[. \\
r & \mapsto &]-\infty, +\infty[.
\end{array}
$$

The evaluation of the first statement constrains the values of x depending on the active branch.

At line 4, we have

$$
\begin{array}{llll}
x & \mapsto & [0, +\infty[& y & \mapsto &]-\infty, +\infty[\\
xm & \mapsto &]-\infty, +\infty[& ym & \mapsto &]-\infty, +\infty[\\
r & \mapsto &]-\infty, +\infty[
\end{array}
$$

while at line 6 we have

$$
\begin{array}{llll}
x & \mapsto &]-\infty, 0[& y & \mapsto &]-\infty, +\infty[\\
xm & \mapsto &]-\infty, +\infty[& ym & \mapsto &]-\infty, +\infty[\\
r & \mapsto &]-\infty, +\infty[.
\end{array}
$$

After the assignment of line 4, we obtain for variable xm the interval $[0, +\infty[$. Similarly, after the assignment of line 6, we obtain for variable xm the interval $-]-\infty, 0[=]0, +\infty[$. The computation of the join in the definition of the abstract collecting semantics at program point 7 returns the interval $]0, +\infty[\cup_{\mathcal{M}} [0, +\infty[= [0, +\infty[$ for xm. However, the join of $]-\infty, 0[$ and $[0, +\infty[$ returns the interval $]-\infty, +\infty[$ for variable x.

The abstract evaluation of program points 8 to 15 follows comparable patterns. Note that the condition $xm \geq ym$ does not provide any meaningful information for this interval-based analysis. We eventually obtain the following abstract environment:

$$
\begin{array}{llll}
x & \mapsto &]-\infty, +\infty[& y & \mapsto &]-\infty, +\infty[\\
xm & \mapsto & [0, +\infty[& ym & \mapsto & [0, +\infty[\\
r & \mapsto & [0, +\infty[.
\end{array}
$$

This analysis has been able to obtain the positivity of r without any assumption on the input values. The same analysis can be done by assuming that the initial abstract environment is:

$$x \; \mapsto \; [0,0] \qquad y \; \mapsto \; [0,0].$$

In that case the final abstract environment obtained is:

$$
\begin{array}{llll}
x & \mapsto & [0,0] & \quad y \;\;\; \mapsto \;\;\; [0,0] \\
xm & \mapsto & [0,0] & \quad ym \;\; \mapsto \;\;\; [0,0] \\
r & \mapsto & [0,0].
\end{array}
$$

Note that abstract environments associated to program points 6, 10, and 14 are associated to the empty environment denoting unreachable program points.

2.3   DEDUCTIVE METHODS

Weakest precondition methods are typically designed for imperative languages. A realistic application will reason on the program as outlined in Section 2.2.1.1 but shall also address the following items:

2.3.1   Loops and recursion in programs

While predicate transformation may seem natural in the previous example, it is less obvious in presence of loops in the control flow graph. A sufficient rule to validate annotations, as defined by Floyd or Hoare, could be:

$$\frac{\vdash \{A \wedge b\} c \{A\}}{\vdash \{A\} \; \text{while } b \text{ do } c \{A \wedge \neg b\}} \; .$$

But it is not compatible with the automatic transformation of predicates as performed in weakest precondition computation. Another way to address this issue is to unroll the loop:

$$\text{while b do c} \equiv \text{if b then c; while b do c else skip.}$$

Then

$$
\begin{aligned}
& wp(\text{while } b \text{ do } c, B) \\
\triangleq \; & wp(\text{if } b \text{ then } c; \text{while } b \text{ do } c \text{ else skip}, B) \\
\triangleq \; & b \Rightarrow wp(c, wp(\text{while } b \text{ do } c, B)) \wedge \neg b \Rightarrow B.
\end{aligned}
$$

Let us denote by $W = wp(\text{while } b \text{ do } c, B)$. We can use the loop unfolding to characterize recursively W:

$$W = (b \Rightarrow wp(c, W) \land \neg b \Rightarrow B).$$

Thanks to Tarski fixpoint theorem, considering the partial order induced by logical implication $\Rightarrow$, i.e., $x \sqsubseteq y \triangleq y \Rightarrow x$, and the monotonic definition of W, this fixpoint exists. But this formula is difficult to compute and may not be representable with a finite set of atoms. If characterizable it captures precisely the loop semantics: the most precise loop invariant, the relationship between input and output, preserved by the loop body.

The solution proposed by Dijkstra [6] is to provide, manually, a weaker loop invariant I, i.e., such that $W \Rightarrow I$. The predicate transformation rule is then defined as

$$wp(\text{while } b \text{ do } c, B)$$
$$\triangleq \quad I \land ((I \land b) \Rightarrow wp(c, I)) \land ((I \land \neg b) \Rightarrow B). \tag{2.6}$$

As a result, any occurrence of loop in programs requires the definition of a loop invariant capturing the loop semantics.

Similarly, in order to prove termination, one needs to exhibit a loop variant, a decreasing sequence in a Noetherian relation, also called a well-founded relation. Typical implementations rely on a positive integer-valued function decreasing at each loop iteration.

2.3.2 Memory model and low-level representation

Until now all computations have been performed on a naive imperative language with real data types, without complex data structure, memory allocation, or function calls.

Serious tools such as Frama-C handle all those constructs. Memory issues are a large part of them. Multiple choices could be made to represent the memory: from the simplest being the Hoare model without pointers or aliases, to a bit level representation. The more complex the memory model, the bigger the generated predicate.

Dedicated analyses such as separation logic [40] can be used to detect aliases or guarantee that pointers x and y are separated, easing the later analyses. Tools such as the Verified Software Toolchain (VST) [98] rely on such analysis.

2.3.3 Underlying logic and automatic proof

A last difficulty in realistic implementations is the expressivity and tools associated to the underlying logic. In Frama-C, the annotation language ACSL [148] (ANSI C Specification Language) is extremely rich and enables the definition of predicates or internal data structures in both functional or axiomatic ways. However, for the same concept, e.g., a linked list or a treelike structure, or an integer-valued function computing the size of a data structure, the generated predicate will widely differ and so do the results of the automatic solver to prove the final proof objective.

Efficient use of these techniques requires the understanding of solver capabilities and their efficiency on different kinds of modelings.

2.4   SMT-BASED MODEL-CHECKING

While SMT-based model-checking can be applied at code level, e.g., the SPACER tool [122, 136], most applications are performed on earlier representations of the system, at model level.

In all cases, a logical representation of the denotational semantics is extracted from the model/code f. It can be as a single predicate associating outputs to inputs, or a more axiomatic definition, for example, relying on a set of Horn clauses. In all cases, it characterizes a transition system with inputs In and outputs Out: $(\Sigma, Init \subseteq \Sigma, T)$ where $T(x, y) \equiv [\![f]\!]_{\mathrm{den}}(x) = y$.

When considering functions with side effects, i.e., depending on memory and modifying it through execution, the typical predicate is

$$T(in, out, mem\_pre, mem\_post).$$

We can also define the initial state of the memory with a predicate:

$$Init(mem).$$

These predicates are valid only for values that satisfy the program semantics. In the memoryless example of Fig. 2.1, we have $T(0, 42)$, $T(1, 46)$, $T(2, 46)$ since these values are valid pairs of input/output, but $T(1, 2)$ is false.

For models without complex data structures this encoding can be rather straightforward. In cases of a variety of data types, casts between values, or complex control flow structures, the encoding can be less easy to define.

Once the encoding is available, one can reason about it. When relying on model-based development such as MATLAB Simulink, ANSYS SCADE, or Lustre, it is possible to extract such an encoding. Since all those models denote synchronous dataflow languages, the semantics of a model is the infinite execution of the blocks semantics.

Let us consider a (possibly infinite) trace $s_0 \to \ldots \to s_i \to \ldots$ of such a system. It corresponds to the sequence of inputs $i_0 \to \ldots \to i_i \to \ldots$ and satisfies the following constraints:

$$Init(s_0) \tag{2.7}$$

$$\forall i \geq 0, \exists o_i, \text{ s.t. } T(i_i, o_i, s_i, s_{i+1}) \tag{2.8}$$

generating the sequence of outputs $o_0 \to \ldots \to o_i \to \ldots$.

Most SMT-based model-checking techniques are based on the induction principle: a way to prove a property invariant over reachable states is to show it inductive over such states. Let $P(s)$ be the predicate encoding of this property.

We recall that the induction principle requires:

$$\forall s \in \Sigma,$$
$$Init(s) \Rightarrow P(s) \qquad \text{(base case)} \qquad (2.9)$$

$$\forall s_1, s_2 \in \Sigma,$$
$$P(s_1) \wedge T(s_1, s_2) \Rightarrow P(s_2). \qquad \text{(inductive case)} \qquad (2.10)$$

However, while the property is inductive over reachable states $[\![f]\!]_{\mathrm{coll}}$:

$$\forall s \in \Sigma,$$
$$Init(s) \Rightarrow P(s) \qquad \text{(base case)} \qquad (2.11)$$

$$\forall s_1, s_2 \in \Sigma \cap [\![f]\!]_{\mathrm{coll}},$$
$$P(s_1) \wedge T(s_1, s_2) \Rightarrow P(s_2). \qquad \text{(inductive case)} \qquad (2.12)$$

The same property may not be inductive over some states $s \in \Sigma \setminus [\![f]\!]_{\mathrm{coll}}$. Such a state would correspond to a spurious counter-example: a state s_1 unreachable but satisfying P such that its successor s_2 by the transition system semantics violates P:

$$(P(s_1) \wedge T(s_1, s_2)) \not\Rightarrow P(s_2).$$

Different approaches exist to attempt to address this issue, without guarantees of success since $[\![f]\!]_{\mathrm{coll}}$ is not computable:

1. Replace $[\![f]\!]_{\mathrm{coll}}$ by some other invariant I of reachable states. The inductive case becomes:[1]

$$\forall s_1, s_2 \in \Sigma,$$
$$(P(s_1) \wedge T(s_1, s_2) \wedge I(s_1) \wedge I(s_2)) \Rightarrow P(s_2). \qquad (2.13)$$

2. Improve the quality of the initial s_1 as part of reachable states: impose it to be part of a path of length k of the transition system. In that case, it is also required to update the base case in order to guarantee property P for the first k reachable states:

$$\forall l \leq k, \forall s_0, \ldots, s_l \in \Sigma,$$
$$Init(s_0) \wedge \bigwedge_{0 \leq i \leq l-1} T(s_i, s_{i+1}) \Rightarrow \bigwedge_{0 \leq i \leq l} P(s_i) \qquad (2.14)$$

---

[1] Note that I may not be inductive with respect to T.

$$\forall s_0, \ldots, s_{k+1} \in \Sigma,$$

$$\bigwedge_{0 \leq i \leq k} (P(s_i) \wedge T(s_i, s_{i+1})) \Rightarrow P(s_{k+1}). \tag{2.15}$$

The second approach is known as k-induction and was first proposed for pure propositional properties and systems [36] and then extended to more general systems using SMT [102]. This is typically the algorithm used in formal verifiers in ANSYS SCADE or MATLAB Simulink.

The first approach is quite natural: instead of looking for a general inductive property we focus on a restricted set of states. Multiple methods were proposed: simple patterns instantiation [101], the use of abstract interpretation [109], the use of quantifier elimination [140], or the dynamic synthesis of property-specific invariants in property-directed reachability (PDR/IC3) [107].

As in deductive methods, the efficiency of the analysis depends on the encoding of the properties and the SMT solver abilities to prove the base and inductive cases.

2.5 ABSTRACT INTERPRETATION (OF COLLECTING SEMANTICS)

The abstract interpretation framework proposed by Cousot and Cousot [8] provides a methodology in which analyses of semantics can be easily defined and proved correct. An essential step of that methodology is to characterize the semantics of interest as a fixpoint of a monotonic operator over a complete lattice. We refer the reader to Miné's PhD manuscript for a very good introduction to the theory [49].

For the moment, let us give the following definition.

Definition 2.4 (Abstract Interpretation). *Abstract Interpretation is a constructive and sound theory for the approximation of semantics expressed as fixpoint of monotonic operators in a complete lattice.*

While this formulation may seem unnatural to the newcomer, it is actually a simple step when it comes to collecting semantics. Collecting semantics is the semantics characterizing reachable states of a program or of a dynamical system. We recall that Σ is the set of all states. We are interested in characterizing all reachable states $s \in \Sigma$. Reachable states form a set of states $S \subseteq \Sigma$ and belong to its powerset $S \in \wp(\Sigma)$. We would like to compute the most precise element of $\wp(\Sigma)$ denoting all reachable states.

Any powerset is a complete lattice. It is fitted with a partial order, the set inclusion $\subseteq$; any subset of elements admits a least upper bound, the set union $\cup$, and a greatest lower bound, the set intersection $\cap$. It is fitted with a lowest element $\emptyset$ and a greatest one Σ. Therefore, our element of interest denoting all reachable states, let's call it $\mathfrak{C}$, is one specific element of the complete lattice $\langle \wp(\Sigma), \subseteq, \cup, \cap, \emptyset, \Sigma \rangle$.

When one considers the underlying update function of the analyzed system—the transition relation of a dynamical system, or a function describing how each program point value is computed from its predecessors—it can be defined as an endomorphism of Σ. It maps a state to a new state. Let $f : \Sigma \to \Sigma$ be such a function. Note that this function does not need to be monotonic in any sense. In order to ease the later notations, we will indifferently denote by f the isomorphism of Σ or its lift $f^\uparrow$ to sets of states $\wp(\Sigma)$:

$$f^\uparrow(S) = \{f(s) \mid s \in S\}.$$

Using f, we can derive the monotonic transfer function F of the collecting semantics. A classical definition of F is the endomorphism of $\wp(\Sigma)$ which accumulates states starting from an initial set of states $Init \in \wp(\Sigma)$:

$$
\begin{aligned}
\wp(\Sigma) &\to \wp(\Sigma) \\
S &\mapsto Init \cup f(S).
\end{aligned}
\tag{2.16}
$$

When one applies recursively this function to the empty set, the infimum $\bot$ of the lattice $\langle \wp(\Sigma), \subseteq, \cup, \cap, \emptyset, \Sigma \rangle$, we characterize the following sequence of sets of states:

$$S_0 = F(\bot) \qquad\qquad\qquad\qquad\qquad\qquad\qquad = Init$$
$$S_1 = F(Init) \qquad\qquad\qquad\qquad\qquad\qquad = Init \cup F(Init)$$
$$S_2 = F(Init \cup f(Init)) \qquad\qquad\quad = Init \cup F(Init) \cup F^2(Init)$$

. . .

Theorem 2.5 (Tarski fixpoint theorem). *Let D be a complete lattice $\langle D, \sqsubseteq, \sqcup, \sqcap, \bot, \top \rangle$ and $f : D \to D$ be a monotonic function. Then the set of fixed points of f in D is also a complete lattice; it admits a least (lfp) and a greatest (gfp) fixpoints.*

$$lfp f = \sqcap \{X \mid F(X) \sqsubseteq X\}$$
$$gfp f = \sqcup \{X \mid X \sqsubseteq F(X)\}.$$

Since F is a monotonic operator of $\langle \wp(\Sigma), \subseteq, \cup, \cap, \emptyset, \Sigma \rangle$, by Tarski fixpoint theorem, a least fixpoint exists. It is defined as the smallest postfixpoint. A postfixpoint is an element $X \in \wp(\Sigma)$ such that $F(X) \subseteq X$.

Then our set of reachable states, the collecting semantics, is *exactly* characterized by

$$\mathfrak{C} = lfp_\emptyset F = \inf_{X \in \wp(\Sigma)} \{F(X) \subseteq X\}. \tag{2.17}$$

Furthermore, the set of fixpoints is fitted with a complete lattice structure: it is closed by join and meet; its infimum is the least fixpoint; and its supremum the greatest one.

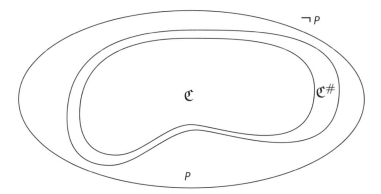

Figure 2.4. Collecting semantics, abstraction, and properties.

2.5.1 Abstracting the fixpoint: fixpoint computation in abstract domains

2.5.1.1 Soundness, incompleteness, and alarms

Despite its proven existence, this exact set of reachable states is very hard to compute in general. The framework of abstract interpretation provides means to abstract it, that is, to compute another value $\mathfrak{C}^{\#}$ of $\wp(\Sigma)$ bigger than $\mathfrak{C}$ for the set inclusion, i.e., containing more states. Some of those states are spurious, they are not reachable in practice, but will be considered as such by the abstraction computed. The validity of a property P is checked with respect to $\mathfrak{C}^{\#}$. P is characterized by the set of states satisfying it: $P = \{s \mid P(s)\}$. In case of success, we have all states in $\mathfrak{C}^{\#}$ satisfy the property, and therefore the subset $\mathfrak{C}$.

$$\mathfrak{C}^{\#} \subseteq P$$
$$\Rightarrow \mathfrak{C} \subseteq P \qquad\qquad \text{by inclusion } \mathfrak{C} \subseteq \mathfrak{C}^{\#}.$$

Figure 2.4 illustrates such inclusions.

In case of failure, one cannot conclude since an erroneous state $s \in \mathfrak{C}^{\#} \setminus P$ could either be in $\mathfrak{C}$ or in spurious states introduced by the abstraction. We speak of an *alarm*. This characterizes the incompleteness of the approach.

2.5.1.2 Abstract domains

An abstraction is meant to approximate sets of states $\wp(\Sigma)$ and is defined by an abstract domain. An abstract domain represents a set of abstract states $D^{\#}$, fitted with a complete lattice structure:[2] $\langle D^{\#}, \sqsubseteq, \sqcup, \sqcap, \bot, \top \rangle$ where $\bot$ and $\top$ denotes infimum and supremum values, respectively.

---

[2]In some cases, such as the one presented in Section 9.3.2, a complete lattice structure is not available. Proofs of convergence are then more complex to achieve.

It also provides means to abstract sets of states $\wp(\Sigma)$ to $D^\#$ and to compute a sound representation as set of states of its elements: those functions are called α and γ, the abstraction and the concretization functions:

$$\alpha : \wp(\Sigma) \to D^\# \qquad\qquad \gamma : D^\# \to \wp(\Sigma).$$

In order to fulfill the abstract interpretation framework methodology, in its most general setting, those abstraction and concretization functions should define a Galois connection:

$$\begin{cases} \text{monotonic } \alpha : \\ \forall s_1, s_2 \in \wp(\Sigma), s_1 \subseteq s_2 \Rightarrow \alpha(s_1) \sqsubseteq \alpha(s_2) \\ \text{monotonic } \gamma : \\ \forall s_1^\#, s_2^\# \in D^\#, s_1 \sqsubseteq s_2 \Rightarrow \gamma(s_1) \subseteq \gamma(s_2) \\ \text{reductivity of } \alpha \circ \gamma : \\ \forall s^\# \in D^\#, \alpha \circ \gamma(s^\#) \sqsubseteq s^\# \\ \text{extensivity of } \gamma \circ \alpha : \\ \forall s \in \wp(\Sigma), s \subseteq \gamma \circ \alpha(s). \end{cases} \qquad (2.18)$$

An abstract domain is also fitted with means to compute, in the abstract, the operations that were performed in the concrete set of states Σ. This ranges from assignments of variables by a linear or polynomial expression, to comparison operations over values, and so on. We denote by $f^\# : D^\# \to D^\#$ the sound abstract counterpart of $f : \wp(\Sigma) \to \wp(\Sigma)$.

2.5.1.3 Soundness in abstract domains

Soundness is guaranteed with respect to the abstraction and concretization functions. We present here the classical definition on a unary operator fun.

$$\begin{aligned} &\forall S \in \wp(\Sigma), S^\# \in D^\#, \\ &S \subseteq \gamma(S^\#) \implies fun(S) \subseteq \gamma(fun^\#(S^\#)). \end{aligned} \qquad (2.19)$$

Soundness could also be expressed relying on α. Intuitively this soundness requirement guarantees that all computations performed in the abstract will, at least, contain the real reachable states and values.

When the abstract domain is defined by a computable Galois connection (α, γ), one can derive automatically these abstract operators such that they compute a sound, yet most precise, solution:

$$op^\#(x) = \alpha \circ op(\gamma(x)). \qquad (2.20)$$

In the case of programs analyzed on their control flow graph representation, such as the ones of Figs. 2.1 and 2.3, (abstract) states of a node with multiple incoming edge, such as a loop head, or an instruction following a conditional statement, are the (abstract) join of the states available in each predecessors.

Using Tarski theorem, one can associate to the concrete set of reachable states $\mathfrak{C}$ the fixpoint of an abstract function $F^\#$:

$$\mathfrak{C}^\# = \mathrm{lfp}_\perp F^\# \tag{2.21}$$

$$= \inf_{X \in D^\#} \left\{ F^\#(X) \sqsubseteq X \right\} \tag{2.22}$$

where $F^\#(S) = \alpha(Init) \sqcup f^\#(S)$.

2.5.1.4   Fixpoint transfer

Thanks to the appropriate choice of α and γ functions, for example, with a Galois connection, and with the additional constraint that the abstraction α commutes with F:

$$\alpha \circ F = F^\# \circ \alpha. \tag{2.23}$$

We have:

$$Init \subseteq \gamma(\alpha(Init)) \qquad \text{(ext. of } \gamma \circ \alpha)$$
$$\Rightarrow \quad F(Init) \subseteq F(\gamma(\alpha(Init))) \qquad \text{(mon. } F)$$
$$\Rightarrow \quad \alpha \circ F(Init) \subseteq \alpha \circ F(\gamma(\alpha(Init))) \qquad \text{(mon. } \alpha)$$
$$\Rightarrow \quad \alpha \circ F(Init) \subseteq F^\# \circ \alpha \circ \gamma(\alpha(Init)) \qquad \text{(using 2.23)}$$
$$\Rightarrow \quad \alpha \circ F(Init) \subseteq F^\#(\alpha(Init)) \qquad \text{(red. of } \alpha \circ \gamma \text{ and mon. of } F^\#$$
$$\text{and } \alpha)$$
$$\Rightarrow \quad \gamma \circ \alpha \circ F(Init) \subseteq \gamma \circ F^\#(\alpha(Init)) \qquad \text{(mon. } \gamma)$$
$$\Rightarrow \quad F(Init) \subseteq \gamma \circ F^\#(\alpha(Init)). \qquad \text{(ext. of } \gamma \circ \alpha)$$

Iterating over F, we obtain

$$\forall n, F^n(Init) \subseteq \gamma \circ F^{\#^n}(\alpha(Init)) \tag{2.24}$$

$$lfp_\emptyset F \subseteq \gamma(lfp_\perp F^\#) \tag{2.25}$$

and therefore

$$\mathfrak{C} \subseteq \gamma(\mathfrak{C}^\#).$$

2.5.2   Effective computation: Kleene iterations and widening

When the abstract domain is fitted with a complete lattice structure,[3] this fixpoint could be accurately computed by Kleene iterations:

$$\mathfrak{C} = lfp_\emptyset F = \lim_{n \to \infty} F^n(\bot). \qquad (2.26)$$

In case of infinite ascending chains of iterates, one relies on so-called *widening* operator to ensure convergence in a finite number of iterations. This operator acts as a rough join operator but has better convergence properties. It is however pessimistic since it introduces numerous spurious states in the abstract representation.

Remark 2.6 (Relative performance).   *In general, SMT-based methods such as MC and DM perform better on disjunctive or integer-based properties. SMT solvers are based on a SAT core and a set of solvers for axiomatized theories. These solvers perform generally well on linear systems and linear properties but have issues with polynomial systems.*

AI typically performs better on the synthesis of numerical invariants because disjunctions are propagated within the abstract representation, using the abstract join $\sqcup$. Abstractions exist that postpone the interpretation of these disjunctions such as partitioned analyses [55, 79], disjunctive completion of domains [9, 19], or delayed join [49] to regain precision but cost too much to be used in a systematic manner.

To summarize, for the most common setting, the effective use of abstract interpretation is the following:

1. Express (collecting) semantics as a fixpoint of a monotonic function F over a lattice of properties. In our case, properties are sets of states.
2. Exhibit an abstract domain for set of states, defining abstraction and concretization functions, lattice operations such as join, and a sound abstract counterpart $F^\#$ of F.
3. Abstract initial states and compute with Kleene iterations the least fixpoint in the abstract.
4. In case of convergence issue, rely on widening to converge to a bigger fixpoint in the abstract.
5. The concretization of this abstract fixpoint is a sound over-approximation of the concrete one.

In practice over-approximation is caused:

---

[3]In theory, it is only required to admit least upper bound for ascending chains. In addition, $F^\#$ should be join complete on these chains, i.e., upper continuous: for all chain $w_0, w_i, \ldots, F^\#(\cup_i w_i) = \cup_i F^\#(w_i)$.

- by the set of properties represented or expressible in the abstract domain (linear relationships, intervals, etc.).
- by the abstraction function and the set of abstract counterparts of concrete functions. For example, in the case of difficult precise definitions of a function such as exp, one can approximate it soundly by a function returning the $\top = \mathbb{R}$ value. While sound, this definition is largely imprecise and will lead to more abstraction when this exp function is used. Another issue appears in presence of nonlinear expression analyzed with an abstract domain restricted to linear properties. In that case the nonlinear expression has to be soundly over-approximated, leading to additional imprecision.
- by the use of widening introducing additional abstraction to the computed element.

2.6   NEED FOR INDUCTIVE INVARIANTS

Basically all formal methods rely on the expression of the property of interest as an inductive invariant over the system semantics. In practice all these techniques benefit from additionally provided invariants. We summarize the use of invariant in the different techniques.

2.6.1   Loop invariants for deductive methods

As mentioned in Section 2.3.1 the analysis of loop with deductive methods requires invariants to be provided to capture the loop semantics. While simple invariants may be easily provided, they may be too weak to capture precisely the loop semantics. For example, the loop invariant expressed in Fig. 2.2 is extremely precise: $R, X, Q \geq 0, Y > 0, X = R + QY$.

If one considers Dijkstra's predicate transformer rule for loops in Eq. (2.6), one can see that the remaining property is essentially I, the invariant provided. Invariants for loops act as the cut-rule in proofs. A sound yet weak invariant will generate a weaker precondition that guarantees the postcondition, but not the weakest. The proof that the provided precondition imply the weaker precondition may be unfeasible.

These loop invariants are either manually provided [147] or computed by other means such as abstract interpretation [83].

2.6.2   Inductive invariants to reinforce transition relation in SMT-based model-checking

Similarly, SMT-based model-checking is essentially based on induction. As mentioned in Section 2.4, different approaches are used to address the lack of availability of the collecting semantics $[\![f]\!]_{\mathrm{coll}}$. While looking different, k-induction, PDR, or invariant injection all amount to the characterization of invariants of $[\![f]\!]_{\mathrm{coll}}$.

2.6.3 Inductive invariants to strengthen abstract interpretation fixpoint computation

Abstract interpretation aims at computing inductive invariants. The definition of abstraction through sound abstract domains enables their composition to improve the analysis results. Since the Cartesian product of two complete lattices is also a complete lattice and since Galois connections can be similarly composed, one can easily define as a sound analysis an analysis that relies on multiple abstractions at the same time. Another interesting construct is the domain reduction: it enables multiple domains to communicate and refine their own (sound) properties. Let us illustrate that notion on a simple example.

Example 2.7. *Consider a set of integer values* $\Sigma = \mathbb{N}$ *and the three following abstractions: sign, interval, and parity.*

$$
\alpha_{sign}(S) = \begin{cases}
\bot_{sign} \ when \ S = \emptyset \\
0 \ when \ S = \{0\} \\
+ \ when \ \forall s \in S, s \geq 0 \\
- \ when \ \forall s \in S, s \leq 0 \\
\top_{sign} \ otherwise.
\end{cases}
$$

$$\alpha_{interval}(S) = (\min S, \max S).$$

$$
\alpha_{parity}(S) = \begin{cases}
\bot_{parity} \ when \ S = \emptyset \\
Odd \ when \ \forall s \in S, s \bmod 2 = 1 \\
Even \ when \ \forall s \in S, s \bmod 2 = 0 \\
\top_{parity} \ otherwise.
\end{cases}
$$

Abstracting a set by the sign of its elements is always less precise than representing more finely the set of values by its lower and upper elements. But those two abstract representations are not comparable with the abstraction that determines whether all values are even or odd.

Abstractions could be combined, however. If both intervals and parity are of interest to us, one can analyze the semantics of the program with both abstractions in the same computation and represent more precisely the interval and parity associated to the abstract set of values. This could lead to further improvements. For example, an interval abstraction may have identified a set $[1, 1000]$ *of reachable values while the parity abstraction guarantees that all values are odd. In that case the interval representation can be refined into* $[1, 9999]$.

One of the major applications of abstract interpretation is the tool Astrée that was designed specifically for the analysis of the Airbus A3XX family control command systems. It combines numerous abstract domains with complex reductions [66]. One of these domains is specifically focused on second-order linear filters in order to bound their reachable states [47].

Chapter Three

Control Systems

THE FOCUS of this book is on control systems. We sketch here their typical development and refer the reader to classical books such as Åström/Murray book [76] or Levine's control handbook [27] for more details on control system design.

Historically, control design started in the continuous world: a system had to be controlled, and its dynamics was captured by the equations of physics, for example, using ordinary differential equations (ODE). Then, control theory provides means to build a controller: another system, used in combination with the system to be controlled, is able to move the system to the requested state.

Figure 3.1 presents a typical process leading to the development of a controller in the aerospace domain. We now give an idea of each step.

3.1 CONTROLLERS' DEVELOPMENT PROCESS

Let us give a naive yet representative process leading to the definition of a control system.

SYSTEM DYNAMICS

At first an identification phase is required to obtain the *plant* dynamics. This identification phase can be complex and relies on various means to describe the system dynamics: a finite element structural model relying on a precise modeling of the aircraft shape, or a rough point mass system with a given number of degree of liberty. For example, for a system like an aircraft able to move in a volume, one can characterize roughly its dynamics by 12 equations defining its position and velocity in an orthogonal basis as well as its angle and angular velocity along the three Euler angles (Yaw, Pitch, Roll).

Typically, one characterizes the sum of forces applied to the system (gravity, thrust, lift, and drag) as we learn in high school. This set of constraints defines the differential equations capturing the dynamics of the system.

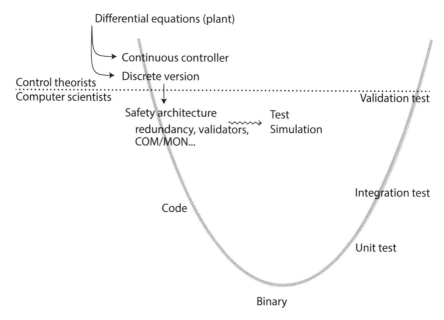

Figure 3.1. Current development process.

LINEARIZATION–TRANSFER FUNCTIONS

LINEARIZATION–TRANSFER FUNCTIONS

Controlling nonlinear dynamics is still an active area of research. In practice, in the conservative aerospace industry, most basic controllers are still defined with old-school linear methods. For these methods the dynamics has to be linear. Since linearized systems are not fully representative far away from the linearization point, multiple of such points are defined, leading to multiple linearized versions of the dynamics. This can be done using Taylor expansion, for example. The general ODE can then be expressed, locally, as a linear differential equation (LDE) expressed over single input and single output signals. The dynamics described by this LDE can be interpreted as a function mapping this input signal $x(t)$ to the output one $y(t)$. In this continuous setting, one defines this function as the linear mapping relating the Laplace transforms of $x(t)$ and $y(t)$. The transfer function is expressed to map those two Laplace transforms.

Control design then provides tools to build a feedback controller: another transfer function which, when associated to the initial transfer function, provides the expected behavior. Various techniques exist to synthesize such a controller: proportional, lead-lag, proportional-integral-derivative (PID), etc.

ANALYSIS

The produced controller can be evaluated with respect to control-level properties. A controller drives the plant in the desired state by minimizing the error between the controller command and the current plant state. This feedback system, the closed-loop system, is analyzed with respect to stability, robustness, and performance. Stability and robustness capture the damping of the system, its ability to converge to goal even in the presence of noise in the feedback loop. Performance evaluates the speed of convergence and the shape of the feedback response (overshoot, number of oscillations, settling time, etc.).

DISCRETIZATION

This controller is meant to be embedded in an onboard computer and to interact with the system sensors and actuators. Depending on the speed of each of these devices, and the available computing resources, an appropriate rate of discretization is chosen. For example, a typical control law for an aircraft runs at 100Hz. But a trajectory planning controller may run at a much lower speed.

COMPLETE CONTROLLERS

Once a discrete controller has been obtained for a linearized version of the plant, a more global one is obtained by combining local controllers. One of the approaches is to synthesize a controller for each linearization point while keeping the same controller synthesis method. Since the previous steps characterized single input single output (SISO) subcontrollers, it is easy to switch the controller depending on the input value. When considering an intermediate value between two linearizations, one can characterize the linear interpolation of controller gains, the coefficients synthesized for each local controller.

Moreover, additional constructs are introduced to account for divergence of integrators in case of a break in the closed-loop system. Saturations or anti-windups act as such and enable the output to remain within given bounds.

INTEGRATION: SAFETY ARCHITECTURE

In critical applications the controller will not be directly embedded on the target platform but rather used in conjunction with a safety architecture used to obtain a fault tolerant system.

This safety architecture is usually identified before the design of the controller itself since it identifies early in the process development the potential causes of failure and their impact on the various systems. These failures can range from faulty parts such as sensors generating false data, a transient error such as a single event upset (SEU) or a multiple bits upset (MBU), or a bug in a software.

This leads to local impacts at the control level with the fusion of input data in case of redundancy in the sensors: validators, alarm detection, voters, etc. The alarm detection mechanisms typically check that the read value lies in an expected range and emit different kinds of alarm signals when a value outside the legal scope is detected.

Figure 3.2 presents an example of such architecture with triplicated input sensors.

At the system level, more complex safety patterns allow the execution of the controller in a distributed fashion, on multiple computers. These different computers

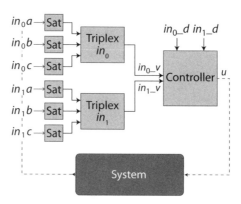

Figure 3.2. Example of a controller with two triplicated inputs.

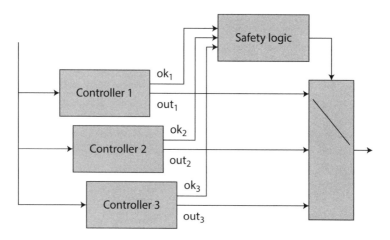

Figure 3.3. Triplication of the controller.

may also run different implementations, to account for hardware (CPU or RAM) and software errors. For example, a first pattern can sequence redundant implementations with only a single one in control as shown in Figure 3.3. Another one, called COM/MON for Command/Monitor, is based on the notion of computer-local observers that detect whether the current output is valid or not. In case of local failure the primary computer leaves the command to the secondary one.

CODE GENERATION AND V&V

Last, once the complete design has been done, the final code is created. It can be automatically generated from model description, or directly coded in C code.

This code is very specific to the control system. It consists mainly of an endless loop, acquiring input data, performing one step of computation, propagating orders to actuators, and awaiting the next clock tick.

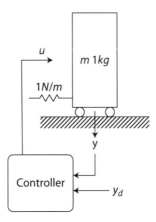

Figure 3.4. Motivating example: a spring-mass damper.

```
while (true) {
in = read_sensors (); // read input data
*state = ctl(*state, in);
actuators(*state); // send orders
wait_next_tick();
}
```

At the verification level, in addition to functional requirements such as the validity of the safety architecture or the alarm mechanism, one needs to prove that the generated code will satisfy the timing constraints imposed by the discretization, as well as prove the absence of runtime errors, such as overflows, that will impact drastically the global behavior of the controlled system, as it happened in the failure of the first Ariane 5 flight [149].

3.2 A SIMPLE LINEAR SYSTEM: SPRING-MASS DAMPER

When considering linear systems, plant and controller, we reuse the running example of [16, 92]. This dynamical system is composed of a single mass and a single spring.

3.2.1 Continuous dynamics: plant and lead-lag controller

First, the plant dynamics is characterized by the following ODE:

$$\frac{d}{dt}x_p = \begin{bmatrix} 0 & 1 \\ -1 & 0 \end{bmatrix} x_p + \begin{bmatrix} 0 \\ 1 \end{bmatrix} u \qquad (3.1)$$

where x_p denotes $\begin{bmatrix} z & \dot z \end{bmatrix}$ the position and velocity of the mass with respect to the origin. The sensor of the plant provides the position z.

The control is performed by a lead-lag controller obtained through classical control recipes where the input y_c is defined as the saturation in the interval $[-1, 1]$ of $y - y_d$ with y the measure of the mass position and $|y_d| \leq 0.5$ a bounded command.

The transfer function of the synthesized controller is:

$$u(s) = -128 \cdot \frac{s+1}{s+0.1} \cdot \frac{s/5+1}{s/50+1} y_c(s). \tag{3.2}$$

The transfer can be expressed as a continuous linear controller using a realization[1] of the above transfer function:

$$\frac{d}{dt} x_c = \begin{bmatrix} -50.1 & 5.0 \\ 1.0 & 0.0 \end{bmatrix} x_c + \begin{bmatrix} 100 \\ 0 \end{bmatrix} \mathrm{SAT}(y_k - y_k^d) \tag{3.3}$$

$$u = \begin{bmatrix} 564.48 & 0 \end{bmatrix} x_c - 1280.$$

3.2.2   Discrete plant dynamics

When producing the embedded controller, the continuous model is discretized at a given rate of execution. This leads to embedded runtime systems which are executed on a platform at the given rate. The rate is chosen according to both the requirements in terms of hardware—one cannot run heavy computation at 1GHz—and in terms of performance—a controller feedback every second may be too slow to control an unstable system such as an inverted pendulum. The typical rate to maintain aircraft stability is 100Hz.

In order to enable the later analyses, we also provide a discretized version of the plant dynamics. Both controller and plant have been discretized at an execution rate of 100Hz.

The plant is described by a linear system over the state variables $p = [x_{p1} \, x_{p2}]^\mathsf{T} \in \mathbb{R}^2$, characterized by the matrices $A_P \in \mathbb{R}^{2 \times 2}$, $B_P \in \mathbb{R}^{1 \times 2}$, and $C_P \in \mathbb{R}^{2 \times 1}$ where u denotes the actuator command of the plant and y the projection of the plant state p over the y sensor:

$$\begin{aligned} p_{k+1} &= A_P p_k + B_P u_k \\ y_{k+1} &= C_P p_{k+1} \end{aligned} \tag{3.4}$$

with

$$A_P := \begin{bmatrix} 1 & 0.01 \\ -0.01 & 1 \end{bmatrix} \quad B_P := \begin{bmatrix} 0.00005 \\ 0.01 \end{bmatrix} \quad C_P := \begin{bmatrix} 1 & 0 \end{bmatrix}.$$

---

[1] This is explained with more details in chapter 7.

3.2.3 Discrete controller dynamics

The controller without saturation is similarly described by a linear system over the state variables $c = [x_{c1}\ x_{c2}]^{\mathsf{T}} \in \mathbb{R}^2$, controlled by both the feedback from the plant sensors $y \in \mathbb{R}^{d_y}$ and the user command $y_d \in \mathbb{R}$, and parametrized by the four real matrices $A_C \in \mathbb{R}^{2 \times 2}$, $B_C \in \mathbb{R}^{1 \times 2}$, $C_C \in \mathbb{R}^{2 \times 1}$, and $D_C \in \mathbb{R}$:

$$
\begin{aligned}
c_{k+1} &= A_C c_k + B_C(y_k - y_{d,k}) \\
u_{k+1} &= C_C c_{k+1} + D_C(y_{k+1} - y_{d,k+1})
\end{aligned}
\tag{3.5}
$$

with

$$
A_C := \begin{bmatrix} 0.4990 & -0.05 \\ 0.01 & 1 \end{bmatrix} \quad B_C := \begin{bmatrix} 1 \\ 0 \end{bmatrix}
$$
$$
C_C := \begin{bmatrix} 564.48 & 0 \end{bmatrix} \quad D_C := -1280.
$$

These numerical values have been obtained by a first-order Euler discretization of the continuous controller.

3.2.4 Closed-loop system

The closed-loop system can be characterized and evaluated. Fig. 3.5 presents the impulse and step response of the closed-loop system.

Let us first consider a version of the closed-loop system, without saturation:

3.2.4.1 System without saturation

The resulting closed-loop system is defined by considering Equations (3.4) and (3.5) at once. It can be expressed over the state space $x := [c\,p]^{\mathsf{T}}$ as

$$
x_{k+1} = Ax_k + By_{d,k}
\tag{3.6}
$$

with

$$
A := \begin{bmatrix} A_C & B_c C_P \\ B_P C_C & A_P + B_P D_C C_P \end{bmatrix}
$$
$$
= \begin{bmatrix}
0.499 & -0.05 & 1 & 0 \\
0.01 & 1 & 0 & 0 \\
0.028224 & 0 & 0.936 & 0.01 \\
5.6448 & 0 & -12.81 & 1
\end{bmatrix}
$$
$$
B := \begin{bmatrix} -B_C \\ -B_P D_C \end{bmatrix} = \begin{bmatrix} -1 \\ 0 \\ 0.064 \\ 12.8 \end{bmatrix}.
$$

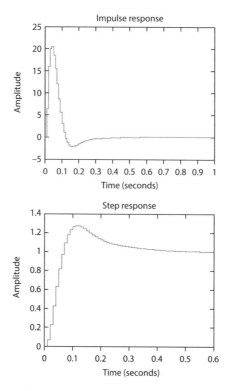

Figure 3.5. Impulse and step response for the controlled spring-mass damper.

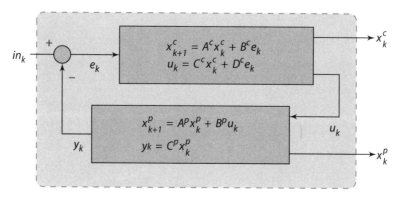

Figure 3.6. Closed-loop system.

```
xc1 = xc2 = xp1 = xp2 = 0;
while (1) {
  yd = acquire_input();
  assert(yd >= -0.5 && yd <= 0.5);
  oxc1 = xc1; oxc2 = xc2; oxp1 = xp1; oxp2 = xp2;
  xc1 = 0.499 * oxc1 - 0.05 * oxc2 + (oxp1 - yd);
  xc2 = 0.01 * oxc1 + oxc2;
  xp1 = 0.028224 * oxc1 + oxp1 + 0.01 * oxp2
        - 0.064 * (oxp1 - yd);
  xp2 = 5.6448 * oxc1 - 0.01 * oxp1 + oxp2
        - 12.8 * (oxp1 - yd);
  wait_next_clock_tick();
}
```

Figure 3.7. Code for the closed-loop system.

$$x_{c1} := 0$$
$$x_{c2} := 0$$
$true$, $x_{p1} := 0$ $-0.5 \le y_d \le 0.5$, $x_{c1} := 0.499\, x_{c1} - 0.05\, x_{c2} + x_{p1} - y_d$
$$x_{p2} := 0$$ $x_{c2} := 0.01\, x_{c1} + x_{c2}$
(s. t.) $\longrightarrow$ (1) $\longleftarrow \bigcirc$ $x_{p1} := 0.028224\, x_{c1} + x_{p1} + 0.01\, x_{p2} - 0.064\,(x_{p1} - y_d)$
                                                            $x_{p2} := 5.6448\, x_{c1} - 0.01\, x_{p1} + x_{p2} - 12.8\,(x_{p1} - y_d)$

Figure 3.8. Control flow graph for code of Figure 3.7.

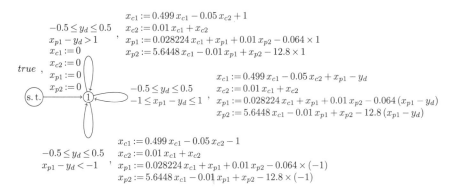

Figure 3.9. Control flow graph for the system with a saturation.

From that formalization, it is possible to characterize a virtual implementation of the closed system as a program. Figure 3.7 displays such code. A control flow graph analysis such as a Kleene based graph reconstruction abstract domain [129] can extract the associated system representation of Figure 3.8.

Remark 3.1. *This corresponds to the system presented in Equation (3.6) with the input y_d bounded by 0.5 ($|y_{d,k}| \leq 0.5$ for all k).*

3.2.4.2 System with saturation

Similarly, the more realistic setting integrating the saturation over $(y - y_d)$ will be defined by the system:

$$x_{k+1} = A x_k + B \operatorname{SAT}(C x_k - y_{d,k}) \tag{3.7}$$

where

$$A := \begin{bmatrix} A_C & 0 \\ B_P C_C & A_P \end{bmatrix} = \begin{bmatrix} 0.499 & -0.05 & 0 & 0 \\ 0.01 & 1 & 0 & 0 \\ 0.028224 & 0 & 1 & 0.01 \\ 5.6448 & 0 & -0.01 & 1 \end{bmatrix}$$

$$B := \begin{bmatrix} B_C \\ B_P D_C \end{bmatrix} = \begin{bmatrix} 1 \\ 0 \\ -0.064 \\ -12.8 \end{bmatrix} \qquad C := \begin{bmatrix} 0 \\ C_P \end{bmatrix}^\mathsf{T} = \begin{bmatrix} 0 \\ 0 \\ 1 \\ 0 \end{bmatrix}^\mathsf{T}$$

and SAT is defined as

$$\operatorname{SAT}(x) = \begin{cases} -1 & \text{if } x < -1 \\ x & \text{if } -1 \leq x \leq 1 \\ 1 & \text{if } x > 1. \end{cases}$$

The control flow graph of the system is presented in Figure 3.9.

Part II

Invariant Synthesis: Convex-optimization Based Abstract Interpretation

Chapter Four

Definitions—Background

THIS PART FOCUSES on the computation of nonlinear numerical invariants for discrete controllers. As mentioned in the motivation part, controllers are usually designed in a continuous setting and then discretized; in both cases a semantics in the real field is assumed. The semantics of interest, in this context, is then a discrete dynamical system with a real semantics. Then, those simple controllers are combined with simple mechanisms such as switches, interpolation of gains, saturations, anti-windups, etc. In order to be able to analyze systems as complex as the ones embedded in aircraft, we consider semantics including piecewise behaviors.

Regarding the analysis of those semantics, an initial natural approach is to apply abstract interpretation on controller programs. However, a long known result for control people can be stated as: "stable linear controllers admit quadratic Lyapunov functions." Unfortunately, most state-of-the-art abstract domains abstract states through linear properties. Furthermore, the current trend was to compute even weaker abstractions, such as octagons [62], to control the complexity of the analyses and obtain nontrivial results in reasonable time. When manipulating nonlinear abstractions, the classical Kleene based approach to fixpoint computation does not seem to be very efficient or appropriate: nonlinear subspaces were not easily fitted with a lattice structure–in other words a least upper bound operator was not as obvious as it is for finite sets of convex polyhedra or intervals. Following the path of control scientists the approaches supported in this book chose to rely on numerical tools, in this case on convex optimization, to solve the so-called Lyapunov equations.

The current chapter presents the formalisms describing discrete dynamical systems and gives an overview on the convex optimization tools and methods used to compute the analyses.

4.1 DISCRETE DYNAMICAL SYSTEMS

A dynamical system is a typical object used in control systems or in signal processing. In some cases, it is eventually implemented in a program to perform the desired feedback control to a cyber-physical system.

Definition 4.1 (State Space). *Let Σ be the state space, a set of states. A dynamical system computes an infinite sequence of states Σ starting from an initial state $init \in X^{Init} \subseteq \Sigma$. The dynamics of the system is defined by a function $f : \Sigma \to \Sigma$. In some cases, the dynamics is also perturbed—or controlled, depending on the point of view—by an external signal, i.e., sequences of values. Let us call them inputs $u \in X^{In}$. The system map is then defined as $f : \Sigma \times X^{In} \to \Sigma$. Let $\bar{\Sigma}$ be the state-input space defined as $\Sigma \times X^{In}$.*

Definition 4.2 (Trajectory). *A trajectory of the system is defined by an initial state $init \in X^{Init}$ and an infinite sequence of inputs $(u_k)_{k \geq 1} \in X^{In}$:*

$$x_0 = init$$
$$x_{n+1} = f(x_n, u_n).$$

Language-wise, model-based languages such as LUSTRE [18], ANSYS SCADE, or MATLAB Simulink provide primitives to build these dynamical systems or controllers relying on simpler constructs. In terms of programs, such dynamical systems can easily be implemented as a *while true loop* initialized by the initial state and performing the update f. The simplest systems are usually directly coded in the target language, e.g., C code, while more advanced systems are *compiled* through autocoders: LUSTRE compilers, SCADE KCG, or MATLAB Embedded Coder (ex-Real Time Workshop RTW).

Let us sketch a typical implementation: the variable u is being read from an external source, e.g., as a mutable variable or an IO call.

```
x = i;
while true {
    u = read();
    x = f(x, in);
}
```

Most systems perform an action at each computation step. In the case of controllers, the action typically moves some actuators in order to impact the controlled system. This generates an output signal, a sequence of produced values $y \in X^{Out}$. This output is computed by a function $g : \Sigma \times X^{In} \to X^{Out}$.

```
x = i;
while true {
    in = read();
    x = f(x, u);
    y = g(x, u);
}
```

A discrete dynamical system is then defined by the following sets Σ, X^{Init}, X^{In}, X^{Out} and functions f, g.

In the following, we specialize this description depending on the considered sets and functions: linear systems, piecewise linear systems, and polynomial ones.

Again, these descriptions are provided at the model level or could be extracted from the implementation [129]. In order to simplify this extraction phase, or to understand it more easily, we assume without loss of generality that the analyzed programs are written in Static Single Assignment (SSA) form, that is, each variable is initialized at most once.

As a last remark, since we are first interested only in the internal state x of the system, the output part is often neglected.

4.1.1 Linear systems

The simplest systems are composed of a single loop and a linear update. While they could seem overly simple to the non-expert, most controllers are linear, from rocket stabilization controllers to aircraft controllers or satellite attitude and orbital control systems (AOCS).

The basic control literature mentions proportional controllers (P), proportional derivative (PD), or proportional-integral-derivative (PID) ones. In all cases, these are linear controllers. In order to obtain more precision, the *order* of the linear controller, i.e., the size of its state space Σ, could be extended, considering a more complex system.

A linear controller is typically implemented by the following code, for a system with $\Sigma = \mathbb{R}^2$ and $X^{In} = X^{Out} = \mathbb{R}$:

```
                                                    Ⓒ
x0 = i0;
x1 = i1;
while true {
  in = read();
  nx0 = a00 * x0 + a01 * x1 + b00 * in;
  nx1 = a10 * x0 + a11 * x1 + b10 * in;
  y  = c00 * x0 + c10 * x1 + d00 * in;
  x0 = nx0;
  x1 = nx1;
}
```

In all systems, assignments of variables are performed using only *parallel assignments*. At the implementation level, this imposes to keep a copy of the variable values before the final updates $x_i = nx_i$.

4.1.1.1 Definition

In this first setting, a linear system is defined over a system state in Σ, represented as a vector of $\mathbb{R}^d$, with inputs in X^{In}, represented as a vector of $\mathbb{R}^m$, and by a pair of matrices $A \in \mathbb{R}^{d \times d}$ $B \in \mathbb{R}^{d \times m}$. Its output in X^{Out}, represented by a vector of $\mathbb{R}^{out}$, is computed similarly by a pair of matrices $C \in \mathbb{R}^{out \times d}$ and $D \in \mathbb{R}^{out \times m}$.

Such a system is defined by the two functions:

$$\begin{cases} f & : & \mathbb{R}^d \times \mathbb{R}^m \to \mathbb{R}^d \\ & & (x, u) \mapsto Ax + Bu \\ g & : & \mathbb{R}^d \times \mathbb{R}^m \to \mathbb{R}^{out} \\ & & (x, u) \mapsto Cx + Du. \end{cases}$$

4.1.1.2 Linear controller example

Let us consider the following Linear Quadratic Gaussian (LQG) Regulator:

```
double x[3] = {0, 0, 0};
double nx[3];
double in;
while (1) {
  in = acquire_input();
  nx[0] = 0.9379*x[0] - 0.0381*x[1] - 0.0414*x[2]
      + 0.0237*in;
  nx[1] = -0.0404*x[0] + 0.968*x[1] - 0.0179*x[2]
      + 0.0143*in;
  nx[2] = 0.0142*x[0] - 0.0197*x[1] + 0.9823*x[2]
      + 0.0077*in;
  x[0] = nx[0]; x[1] = nx[1]; x[2] = nx[2];
  wait_next_clock_tick();  // a tick every 10 ms
      for instance
}
```

This characterizes the following dynamical system.

Example 4.3 (Linear system example). *Let* $\Sigma = \mathbb{R}^3$, $X^{Init} = \{(0, 0, 0)\}$, *and* $X^{In} = \mathbb{R}$, *with the following matrices* A *and* B:

$$A := \begin{bmatrix} 0.9379 & -0.0381 & -0.0414 \\ -0.0404 & 0.968 & -0.0179 \\ 0.0142 & -0.0197 & 0.9823 \end{bmatrix} B := \begin{bmatrix} 0.0237 \\ 0.0143 \\ 0.0077 \end{bmatrix}.$$

4.1.2 Switched linear systems: constrained piecewise affine discrete-time systems

Most systems are not purely linear. The programs or systems we consider here are composed of a single loop with possibly a complicated switch-case type loop body. Our switch-case loop body is supposed to be written as a nested sequence of *ite* statements, or as a switch:

```
x = i;
while true {
   in = read();
   switch
      c₁  →  x = f₁(x, in);
      c₂  →  x = f₂(x, in);
      c₃  →  x = f₃(x, in);
      _   →  x = f₄(x, in);
}
```

Moreover, we suppose that the analyzed programs are written in affine arithmetic, both the switch conditions c_i and the associated update functions f_i. Consequently, the programs analyzed here can be interpreted as constrained piecewise affine discrete-time systems.

4.1.2.1 Polyhedral partitioning of $\bar{\Sigma}$

The term piecewise affine means that there exists a polyhedral partition $\{X^i, i \in \mathbb{I}\}$ of the state-input space $\bar{\Sigma} \subseteq \mathbb{R}^{d+m}$ such that for all $i \in \mathbb{I}$, the dynamic of the system is affine and represented by the following relation for all $k \in \mathbb{N}$:

$$\text{if } (x_k, u_k) \in X^i, \ x_{k+1} = A^i x_k + B^i u_k + b^i, k \in \mathbb{N} \tag{4.1}$$

where $A^i \in \mathbb{R}^{d \times d}$, $B^i \in \mathbb{R}^{d \times m}$, and $b^i \in \mathbb{R}^d$. As in the linear case, the variable $x \in \mathbb{R}^d$ refers to the state variable and $u \in \mathbb{R}^m$ refers to some input variable.

We define a partition of the state-input space as a family of nonempty sets X^i such that:

$$\bigcup_{i \in \mathbb{I}} X^i = \bar{\Sigma}, \ \forall i, j \in \mathbb{I}, \ i \neq j, X^i \cap X^j = \emptyset. \tag{4.2}$$

In the current setting, since X^i are convex polyhedra, we characterize polyhedral partitions of the state-input space. From now on, we call X^i *cells*.

4.1.2.2   Affine conditions: strict and weak affine convex constraints

Cells $\{X^i\}_{i \in \mathbb{I}}$ are convex polyhedra which can contain both strict and weak inequalities.

Definition 4.4 (Cells as Convex Polyhedra). *Cells can be represented by n_i linear constraints, as a $n_i \times (d+m)$ matrix T^i, and a vector $c^i \in \mathbb{R}^{n_i}$. We denote by $\mathbb{I}_s^i$ the set of indices which represent strict inequalities for the cell X^i, denote by T_s^i and c_s^i the parts of T^i and c^i corresponding to strict inequalities, and by T_w^i and c_w^i the one corresponding to weak inequalities. Finally, we have the matrix representation given by Equation (4.3).*

$$X^i = \left\{ \begin{pmatrix} x \\ u \end{pmatrix} \in \mathbb{R}^{d+m} \,\middle|\, T_s^i \begin{pmatrix} x \\ u \end{pmatrix} \ll c_s^i, \ T_w^i \begin{pmatrix} x \\ u \end{pmatrix} \leq c_w^i \right\}. \tag{4.3}$$

We use the following notations: $y \leq z$ is a partial order built as the piecewise lift of the total order over reals to vectors, meaning that for all coordinates l, $y_l \leq z_l$. The other relation $y \ll z$ is the strict version, meaning that for all coordinates l, $y_l < z_l$.

4.1.2.3   Homogenization: encoding affine system as a linear one

In order to simplify the following analyses, it is easier to consider a linear system rather than an affine one. Therefore, we define a homogeneous flavor of the system dynamics: instead of considering a system state in $\mathbb{R}^d$ with inputs in $\mathbb{R}^m$, we manipulate system states in $\mathbb{R}^{1+d+m}$. We will need homogeneous versions of update functions and thus introduce the $(1+d+m) \times (1+d+m)$ matrices F^i defined as follows:

$$F^i = \begin{pmatrix} 1 & 0_{1 \times d} & 0_{1 \times m} \\ b^i & A^i & B^i \\ 0 & 0_{m \times d} & \mathrm{Id}_{m \times m} \end{pmatrix} \tag{4.4}$$

where $\mathrm{Id}_{m \times m}$ denotes the identity matrix of dimension $m \times m$.

The system defined in Equation (4.1) can be rewritten as $(1, x_{k+1}, u_{k+1})^\intercal = F^i(1, x_{k+1}, u_k)^\intercal$. Note that, in order to obtain a square matrix, we introduce a "virtual" dynamic law $u_{k+1} = u_k$ on the input variable in Equation (4.4). It will not be used in the following analyses.

Let $p = card(\mathbb{I})$, the global system can be defined as:

$$\begin{cases} (1, x_{k+1}, u_{k+1})^\mathsf{T} = F^1(1, x_k, u_k)^\mathsf{T} & \text{w. } (x_k, u_k) \in X^1 \\ (1, x_{k+1}, u_{k+1})^\mathsf{T} = F^2(1, x_k, u_k)^\mathsf{T} & \text{w. } (x_k, u_k) \in X^2 \\ \ldots \\ (1, x_{k+1}, u_{k+1})^\mathsf{T} = F^p(1, x_k, u_k)^\mathsf{T} & \text{w. } (x_k, u_k) \in X^p. \end{cases} \qquad (4.5)$$

4.1.2.4 Piecewise linear discrete dynamical system example

Let us consider the following program. It is constituted by a single while loop with two nested conditional branches in the loop body, characterizing four cells.

```
                                        Ⓒ
(x,y)∈[-9,9]×[-9,9];
while(true)
ox=x;
oy=y;
read(u);  \\u∈[-3,3]
if (-9*ox+7*y+6*u<5){
  if(-4*ox+8*oy-8*u<4){
    x=0.4217*ox+0.1077*oy+0.5661*u;
    y=0.1162*ox+0.2785*oy+0.2235*u-1;
  }
  else {  \\4*ox-8*oy+8*u<-4
    x=0.4763*ox+0.0145*oy+0.9033*u;
    y=0.1315*ox+0.3291*oy+0.1459*u+9;
  }
}
else {  \\9*ox-7*y-6*u<-5
  if(-4*ox+8*oy-8*u<4){
    x=0.2618*ox+0.1107*oy+0.0868*u-4;
    y=0.4014*ox+0.4161*oy+0.6320*u+4;
  }
  else {  \\4*ox-8*oy+8*u<-4
    x=0.3874*ox+0.00771*oy+0.5153*u+10;
    y=0.2430*ox+0.4028*oy+0.4790*u+7;
  }
}
```

The initial condition of the piecewise affine system is $(x, y) \in [-9, 9] \times [-9, 9]$ and the polytope where the input variable u lives is $\mathcal{U} = [-3, 3]$.

We can rewrite this program as a piecewise affine discrete-time dynamical system using our notations. We give details on the matrices T_s^i and T_w^i and vectors c_s^i and c_w^i (see Equation (4.3)) which characterize the cells and on the matrices F^i representing the homogeneous version (see Equation (4.4)) of affine laws in the cell X^i.

Example 4.5 (Piecewise linear system example). Let $\Sigma = \mathbb{R}^2$, $X^{Init} = [-9, 9] \times [-9, 9]$, $X^{In} = [-3, 3]$, $card(\mathbb{I}) = 4$ with the following matrices and vectors:

$$F^1 = \begin{pmatrix} 1 & 0 & 0 & 0 \\ 0 & 0.4217 & 0.1077 & 0.5661 \\ -1 & 0.1162 & 0.2785 & 0.2235 \\ 0 & 0 & 0 & 1 \end{pmatrix},$$

$$\begin{cases} T_s^1 = \begin{pmatrix} -9 & 7 & 6 \\ -4 & 8 & -8 \end{pmatrix}, \\ c_s^1 = (5\ 4)^\mathsf{T} \end{cases}$$

$$\begin{cases} T_w^1 = \begin{pmatrix} 0 & 0 & 1 \\ 0 & 0 & -1 \end{pmatrix} \\ c_w^1 = (3\ 3)^\mathsf{T} \end{cases}$$

$$F^2 = \begin{pmatrix} 1 & 0 & 0 & 0 \\ 0 & 0.4763 & 0.0145 & 0.9033 \\ 9 & 0.1315 & 0.3291 & 0.1459 \\ 0 & 0 & 0 & 1 \end{pmatrix},$$

$$\begin{cases} T_s^2 = (-9 \quad 7 \quad 6) \\ c_s^2 = 5 \end{cases},$$

$$\begin{cases} T_w^2 = \begin{pmatrix} 4 & -8 & 8 \\ 0 & 0 & 1 \\ 0 & 0 & -1 \end{pmatrix} \\ c_w^2 = (-4\ 3\ 3)^\mathsf{T} \end{cases}$$

$$F^3 = \begin{pmatrix} 1 & 0 & 0 & 0 \\ -4 & 0.2618 & 0.1177 & 0.0868 \\ 4 & 0.4014 & 0.4161 & 0.6320 \\ 0 & 0 & 0 & 1 \end{pmatrix},$$

$$\begin{cases} T_s^3 = (-4 \quad 8 \quad -8) \\ c_s^3 = 4 \end{cases},$$

$$\begin{cases} T_w^3 = \begin{pmatrix} 9 & -7 & -6 \\ 0 & 0 & 1 \\ 0 & 0 & -1 \end{pmatrix} \\ c_w^2 = (-5\ 3\ 3)^\mathsf{T} \end{cases}$$

$$F^4 = \begin{pmatrix} 1 & 0 & 0 & 0 \\ 10 & 0.3874 & 0.0771 & 0.5153 \\ 7 & 0.2430 & 0.4028 & 0.4790 \\ 0 & 0 & 0 & 1 \end{pmatrix},$$

$$\begin{cases} T_w^4 = \begin{pmatrix} 9 & -7 & -6 \\ 4 & -8 & 8 \\ 0 & 0 & 1 \\ 0 & 0 & -1 \end{pmatrix} \\ c_w^4 = (-5 \ -4 \ 3 \ 3)^{\mathsf{T}}. \end{cases}$$

4.1.3 Piecewise polynomial systems

A last flavor of considered systems is the further extension to polynomial constraints and updates: piecewise polynomial discrete-time dynamical systems. Let us first recall some definitions of polynomial functions in $\mathbb{R}^d$.

Definition 4.6 (Polynomial Functions of $\mathbb{R}^d$). *A function f from $\mathbb{R}^d$ to $\mathbb{R}$ is a polynomial if and only if there exists $k \in \mathbb{N}$, a family $\{c_\alpha \mid \alpha = (\alpha_1, \ldots, \alpha_d) \in \mathbb{N}^d, |\alpha| = \alpha_1 + \ldots + \alpha_d \le k\}$ such that for all $x \in \mathbb{R}^d$, $f(x) = \sum_{|\alpha| \le k} c_\alpha x_1^{\alpha_1} \ldots x_d^{\alpha_d}$. By extension a function $f : \mathbb{R}^d \mapsto \mathbb{R}^d$ is polynomial if and only if all its coordinate functions are polynomials. Let $\mathbb{R}[x]$ stands for the set of d-variate polynomials.*

We focus now on programs composed of a single loop with a possibly complicated switch-case type loop body.

```
x ∈ X^Init ;
while true {
    case  (r₁¹(x)<#0 and  ...  and  r¹_{n₁}(x)<#0) :
              x = T¹(x) ;
    case  ...
    case  (r_i¹(x)<#0 and  ...  and  r¹_{n_i}(x)<#0) :
              x = T^i(x) ;
}
```

4.1.3.1 Basic semialgebraic set

In this setting, conditions are expressed as a conjunction of weak polynomial inequalities $r(x) \le 0$ or strict polynomial inequalities $r(x) < 0$. These functions, describing guards, are real-valued polynomials of the state-input space: $\bar{\Sigma} \to \mathbb{R}$. Such conditions characterize a *basic semialgebraic set*. Recall that a set $C \subseteq \mathbb{R}^d$ is said to be a *basic semialgebraic set* if there exist $g_1, \ldots, g_m \in \mathbb{R}[x]$ such that

$C = \{x \in \mathbb{R}^d \mid g_j(x) <^\# 0, \forall\, j = 1, \dots, m\}$, where $<^\#$ is used to encode either a strict $<$ or a weak $\leq$ inequality.

4.1.3.2 Semialgebraic partitioning of $\bar{\Sigma}$

These basic semialgebraic sets of $\bar{\Sigma} = \mathbb{R}^{d+m}$ act as the cells we used in piecewise affine systems. They characterize a partition of the state-input space $\bar{\Sigma}$. Let $\mathbb{I}$ be the set of basic semialgebraic cells X^i.

$$X^i = \left\{ x \in \mathbb{R}^{d+m} \;\middle|\; \bigwedge_{1 \leq j \leq ns_i} r^i_{s_j} < 0 \; \bigwedge_{1 \leq j \leq nw_i} r^i_{w_j} \leq 0 \right\} \tag{4.6}$$

where ns_i and nw_i denote, respectively, the number of strong and weak polynomial constraints in the semialgebraic set X^i.

Cells X^i satisfy Eq. (4.2): they form a semialgebraic partition of $\bar{\Sigma}$. As a result, any element of $\bar{\Sigma}$ belongs to exactly one cell X^i.

4.1.3.3 Polynomial updates

Assignments associated to each cell X^i with $i \in \mathbb{I}$ are defined by polynomial functions T^i.

$$\text{if } x_k \in X^i, \; x_{k+1} = T^i(x_k). \tag{4.7}$$

For systems without input, we have T^i a d-variate polynomial: $T^i \in \mathbb{R}[x]$; in case of systems with input in $\mathbb{R}^m$, T^i is a polynomial function $T^i \in \mathbb{R}^{d+m} \to \mathbb{R}^d$, where each coordinate function is a polynomial $\mathbb{R}^{d+m} \to \mathbb{R}^d$.

4.1.3.4 Definition of a piecewise polynomial system (PPS)

We assume that $\mathbb{I}$ is finite and that the initial condition x_0 belongs to some compact basic semialgebraic set X^{Init} satisfying Eq. (4.6). For the program, X^{Init} is the set in which the variables are supposed to be initialized.

Definition 4.7 (Piecewise Polynomial System). *A constrained polynomial piecewise discrete-time dynamical system (PPS) is the quadruple* $(X^{Init}, X^{In}, \bar{\Sigma}, \mathcal{L})$ *with:*

- $X^{Init} \subseteq \Sigma \subseteq \mathbb{R}^d$ *is the compact basic semialgebraic set of the possible initial conditions;*
- $X^{In} \subseteq \mathbb{R}^m$ *is the basic semialgebraic set where the input variable lives;*
- $\bar{\Sigma} := \{X^i, i \in \mathbb{I}\}$ *is a partition as defined in Equation (4.2);*
- $\mathcal{L} := \{T^i, i \in \mathcal{I}\}$ *is the family of the polynomials from $\mathbb{R}^{d+m}$ to $\mathbb{R}^d$, w.r.t. the partition $\bar{\Sigma}$ satisfying Equation (4.7).*

4.1.3.5 Piecewise polynomial system example

From now on, we associate a PPS representation to each program of the form described earlier.

Let us consider a concrete example. The program below involves four variables and contains an infinite loop with a conditional branch in the loop body. Each branch update is defined by a polynomial function. The parameters c_{ij} (resp. d_{ij}) are given parameters. During the analysis, we only keep the variables x_1 and x_2 since $oldx_1$ and $oldx_2$ are just memories.

```
                                                    Ⓒ
x₁, x₂ ∈ [a₁, a₂] × [b₁, b₂];
oldx₁ = x₁;
oldx₂ = x₂;
while (-1 <= 0){
    oldx₁ = x₁;
    oldx₂ = x₂;
    case :  oldx₁^2 + oldx₂^2 <= 1 :
        x₁ = c₁₁ * oldx₁^2 + c₁₁ * oldx₂^3;
        x₂ = c₂₁ * oldx₁^3 + c₂₂ * oldx₂^2;
    case :   -oldx₁^2 - oldx₂^2 < -1
        x₁ = d₁₁ * oldx₁^3 + d₁₂ * oldx₂^2;
        x₂ = d₂₁ * oldx₁^2 + d₂₂ * oldx₂^2;
}
```

Example 4.8 (Piecewise polynomial system example). *The associated PPS corresponds to the input-empty quadruple* $(X^{Init}, \emptyset, \{X^1, X^2\}, \{T^1, T^2\})$. *In this case* $\bar{\Sigma} = \Sigma$. *We have the set of initial conditions:*

$$X^{Init} = [a_1, a_2] \times [b_1, b_2],$$

the partition verifying Equation (4.2) is:

$$X^1 = \{x \in \mathbb{R}^2 \mid x_1^2 + x_2^2 \le 1\},$$
$$X^2 = \{x \in \mathbb{R}^2 \mid -x_1^2 - x_2^2 < -1\},$$

and the polynomials relative to the partition $\{X^1, X^2\}$ *are:*

$$T^1(x) = \begin{pmatrix} c_{11}x_1^2 + c_{12}x_2^3 \\ c_{21}x_1^3 + c_{22}x_2^2 \end{pmatrix}$$

and

$$T^2(x) = \begin{pmatrix} d_{11}x_1^3 + d_{12}x_2^2 \\ d_{21}x_1^2 + d_{22}x_2^2 \end{pmatrix}.$$

4.2 ELEMENTS OF (APPLIED) CONVEX OPTIMIZATION

This section intends to provide elements to the computer scientist to understand basic principles of convex optimization and the typical approaches to manipulate such optimization problems. We refer the interested reader to the excellent book *Convex Optimization* by Boyd and Vandenberghe [46] for a more thorough introduction. Other valuable references include *Numerical Optimization* by Nocedal and Wright [63], and *Éléments d'optimisation diffrentiable* by Gilbert [120].

In the following we focus on optimization problems of the form

$$\begin{array}{ll} \min & f_0(x) \\ \text{s.t.} & f_i(x) \leq b_i \text{ for } i \in [1, m]. \end{array} \tag{4.8}$$

Here $x \in \mathbb{R}^n$ is the optimization variable. $f_0 \in R^n \to \mathbb{R}$ denotes the objective function and the functions $f_i \in R^n \to \mathbb{R}$, with associated bound b_i, the constraints.

A solution x of (4.8) is feasible if it satisfies all constraints. It is optimal if it is the smallest of all feasible ones.

An optimization algorithm is a numerical tool that computes or approximates such feasible optimal solutions.

4.2.1 Convex conic optimization

In the case where the only part of the problem is the objective function, i.e., no constraint is provided, then classical methods such as gradient, conjugate gradient, or Newton methods will iteratively approximate the solution. These algorithms compute a sequence of points of $\mathbb{R}^n$ by updating the previous point with a local descent direction d_k obtained by considering the derivative of the objective function (aka the gradient).

$$x_{k+1} = x_k + \alpha_k d_k.$$

Here α_k denotes the step size. Both α_k and d_k depend on the current point x_k and typically rely on $f_0'(x_k)$ (aka $\nabla f(x_k)$). Kantorovich's theorem characterizes conditions imposed on f_0 to guarantee the existence of a unique solution and the convergence to it for Newton's method. These constraints amount to provide a bound on the variation of the function, its Lipschitz constant.

Solving a general case of optimization problems with constraints is still an open question. However, a solution to guarantee the existence of such bound is to constrain the functions f_0, and $f_i, \forall i \in [1, m]$ to be convex. We recall that a function f is convex when $\forall \alpha, \beta \geq 0, \alpha + \beta = 1, f(\alpha x + \beta y) \leq \alpha f(x) + \beta f(y)$. In that case any local optimal point is also a global optimal one. An even stronger condition would be to require it to be linear, i.e., $f(\alpha x + \beta y) = \alpha f(x) + \beta f(y)$.

Convex optimization is then a restriction of general optimization to the following problem:

$$
\begin{aligned}
\min \quad & f_0(x) \\
\text{s.t.} \quad & f_i(x) \leq 0 \text{ for } i \in [1, m] \\
& a_j^\mathsf{T} x = b_j \text{ for } j \in [1, p]
\end{aligned}
\tag{4.9}
$$

where $f, 0, f_i$ are convex functions. Note that equality constraints have to be affine: they correspond to a conjunction of two convex inequalities: $f(x) \leq 0 \wedge f(x) \geq 0$; the only solution is to require f to be affine or linear.

A well-known case of this convex optimization problem is linear optimization or linear programming, in which a linear objective function f_0 is optimized while satisfying the linear constraints f_i.

This notion of convex optimization can be further extended to more general convex sets: convex cones. A cone $\mathcal{K}$ is a subset of $\mathbb{R}^n$ closed by positive scaling: $\forall x \in \mathcal{K}, \theta \geq 0, \theta x \in \mathcal{K}$. A convex cone satisfies: $\forall x, y \in \mathcal{K}, \theta_1, \theta_2 \geq 0, \theta_1 x + \theta_2 y \in \mathcal{K}$. Such convex cone can be fitted with partial order $\preceq$ such that $\forall x, y \in \mathcal{K}, (x \preceq y) \equiv (y - x \in \mathcal{K})$. By extension a function convex in the cone is $\mathcal{K}$-convex.

In that setting a convex conic optimization problem is defined as

$$
\begin{aligned}
\min \quad & f_0(x) \\
\text{s.t.} \quad & f_i(x) \preceq_{\mathcal{K}} 0 \text{ for } i \in [1, m] \\
& Ax = b
\end{aligned}
\tag{4.10}
$$

where $f_0 \in \mathbb{R}^n \to \mathbb{R}$ is convex, $f_i \in \mathbb{R}^n \to \mathcal{K}$ are $\mathcal{K}$-convex functions, and $A \in \mathbb{R}^{p \times n}$.

As a first specialization, we speak about linear problems when f_0, f_i are linear. Each function f can then be described as a scalar product $\langle \cdot, \cdot \rangle$ when real-valued, or as a product by a matrix when $\exists m, \mathcal{K} \subseteq \mathbb{R}^m$. In the following we denote the function f_0 by a constant vector c: $f_0(x) = \langle c, x \rangle$, and the functions f_i by the pair A_i, b_i such that $f_i(x) = A_i x - b_i$.

$$
\begin{aligned}
\min \quad & \langle c, x \rangle \\
\text{s.t.} \quad & A_i x - b_i \preceq_{\mathcal{K}} 0 \text{ for } i \in [1, m] \\
& Ax = b.
\end{aligned}
\tag{4.11}
$$

Let us now focus on special cases depending on the cone $\mathcal{K}$ considered.

4.2.1.1 Polytopes

When $\mathcal{K} = \mathbb{R}^+$, a famous case is the optimization over closed polyhedra. Each constraint characterizes a subspace of $\mathbb{R}^n$, the feasible set being the intersection of these subspaces, a convex set. The goal is to optimize a linear function over this bounded convex set. In the case of bounded feasible set, a finite number of vertices characterize the polytope. Since the optimal solution is necessarily on a vertex, the simplex method enumerates these vertices and computes the optimal one.

4.2.1.2 Positive semidefinite cone

Let us consider the set $\mathbb{S}^n$ of symmetric matrices of $\mathbb{R}^{n \times n}$:

$$\mathbb{S}^n = \left\{ X \in \mathbb{R}^{x \times n} \,|\, X = X^{\mathsf{T}} \right\}.$$

The set $\mathbb{S}^n_+$ of positive semidefinite matrices is the subset of matrices of $\mathbb{S}^n$ admitting only positive eigenvalues.

$$\mathbb{S}^n_+ = \{ X \in \mathbb{S}^n \,|\, X \succeq 0 \}.$$

Equivalently, we have:

$$\mathbb{S}^n_+ = \{ X \in \mathbb{S}^n \,|\, \forall x \in \mathbb{R}^n, x^{\mathsf{T}} X x \geq 0 \}.$$

This set is a convex cone: it is closed by addition and external multiplication by a positive scalar. Optimizing over this cone leads to a problem of the form:

$$
\begin{aligned}
\min \quad & \langle c, x \rangle \\
\text{s.t.} \quad & \sum_{i \in [1,n]} x_i F_i + G \preceq 0 \\
& Ax = b.
\end{aligned}
\tag{4.12}
$$

Here x is a vector and matrices $G, F_1, \ldots, F_n \in \mathbb{S}^n_+$. The inequality is known as a *Linear Matrix Inequality* (LMI).

Indeed, we can easily have unknown matrices since a matrix $A \in \mathbb{R}^{n \times n}$ can be expressed as $\sum_{i=0,j=0}^{n-1,n-1} A_{i,j} E^{i,j}$, where $E^{i,j}$ is the matrix with zeros everywhere except a one at line i and column j. Likewise, multiple LMIs can be grouped into one since $A \succeq 0 \wedge B \succeq 0$ is equivalent to $\begin{bmatrix} A & 0 \\ 0 & B \end{bmatrix} \succeq 0$.

Efficient solvers for semidefinite programming (SDP), based on interior point method algorithms are available such as Mosek [32], SDPA [97] or CSDP [30]. For more details about SDP, we refer the interested reader to [28].

4.2.1.3 Sum-of-square polynomials

Let $\mathbb{R}[x]$ be the set of multivariate polynomials of $\mathbb{R}^n$ and $\mathbb{R}[x]_{2m}$ its restriction to polynomials of degree at most $2m$. We denote by $\Sigma[x] \subset \mathbb{R}[x]$ the cone of sum-of-squares (SOS) polynomials, that is,

$$\Sigma[x] := \left\{ \sum_i q_i^2, \text{ with } q_i \in \mathbb{R}[x] \right\}. \tag{4.13}$$

The existence of an SOS representation for a given polynomial is an approach to *Positivestellensatz* witness, a sufficient condition to prove its global nonnegativity, i.e., $\forall p(x) \in \Sigma[x], p(x) \geq 0$. The SOS condition (4.13) is equivalent to the existence of a positive semidefinite matrix Q such that

$$p(x) = Z^\mathsf{T}(x)QZ(x) \tag{4.14}$$

where $Z(x)$ is a vector of monomials of degree less than or equal to $deg(p)/2$.

Searching for a positive polynomial of a given degree $d = 2m$ amounts to solving a semidefinite optimization problem and synthesizing the matrix $Q \succeq 0$, satisfying Eq. (4.14).

Example 4.9. *Consider the bi-variate polynomial $q(x) := 1 + x_1^2 - 2x_1x_2 + x_2^2$. With $Z(x) = (1, x_1, x_2)$, one looks for a semidefinite positive matrix Q such that the polynomial equality $q(x) = Z(x)^\mathsf{T} Q Z(x)$ holds for all $x \in \mathbb{R}^2$.*
The matrix

$$Q = \begin{pmatrix} 1 & 0 & 0 \\ 0 & 1 & -1 \\ 0 & -1 & 1 \end{pmatrix}$$

satisfies this equality and has three nonnegative eigenvalues, which are 0, 1, and 2, respectively associated to the three eigenvectors $e_0 := (0, 1, 1)^\mathsf{T}$, $e_1 := (1, 0, 0)^\mathsf{T}$, and $e_2 := (0, -1, 1)^\mathsf{T}$. Defining the matrices $L := (e_1 \, e_2 \, e_0) = \begin{pmatrix} 1 & 0 & 0 \\ 0 & 1 & 1 \\ 0 & -1 & 1 \end{pmatrix}$ and $D = \begin{pmatrix} 1 & 0 & 0 \\ 0 & 2 & 0 \\ 0 & 0 & 0 \end{pmatrix}$, one obtains the decomposition $Q = L D L^{-1}$ and the equality $q(x) = (L \, Z(x)) \, D \, (L^{-1} \, Z(x)) = \sigma(x) = 1 + (x_2 - x_1)^2$, for all $x \in \mathbb{R}^2$. The polynomial σ is called an SOS certificate and guarantees that q is nonnegative.

An SOS optimization problem can be defined as

$$\begin{array}{ll} \min & \langle c, x \rangle \\ \text{s.t.} & p_i(x) \in \Sigma[x] \text{ for } i \in [1, m] \\ & Ax = b. \end{array} \tag{4.15}$$

SOS programming solvers provide a front end, easing the translation of an SOS-based optimization problem into SDP. Each sum-of-square polynomial constraint p_i is associated to a symmetric positive semidefinite matrix Q_i. By identification, coefficients of the matrices Q_i are associated to equality constraints depending on the expression characterizing the polynomial p_i. Once the SDP problem is solved, its solution is used to rebuild the polynomial problem and provides the positive certificate of the SOS problem (4.15).

The MATLAB toolbox YALMIP [48] provides such a front end.

More references regarding SOS-based polynomial optimization can be found in Parrilo [45] and Lasserre [89] works.

4.2.2   Convex optimization tools

When manipulating optimization problems, there are few standard operations that enable us to relax a problem into a solvable one. We present here a few of those techniques.

4.2.2.1   Convexifying constraints: S-Procedure, Lagrangian relaxation

When facing nonconvex constraints such as implication between positive definite matrices

$$P_1 \succeq 0 \implies P_2 \succeq 0$$

we can express a sufficient condition in a convex way. The S-Procedure provides such a relaxation.

Theorem 4.10 (S-Procedure).   *For any* $P, P_1, \ldots, P_k \in \mathbb{R}^{n \times n}$ *and* $b, b_1, \ldots,$ *$b_k \in \mathbb{R}$ and $b, b' \in \mathbb{R}$, the following*

$$\exists \tau_1, \ldots, \tau_k \in \mathbb{R},$$
$$\left(\bigwedge_{i=1}^{k} \tau_i \geq 0 \right) \wedge \begin{bmatrix} -P & 0 \\ 0 & b \end{bmatrix} - \sum_{i=1}^{k} \tau_i \begin{bmatrix} -P_i & 0 \\ 0 & b_i \end{bmatrix} \succeq 0 \qquad (4.16)$$

is a sufficient condition for

$$\forall x \in \mathbb{R}^n, \left(\bigwedge_{i=1}^{k} x^{\mathsf{T}} P_i \, x \leq b_i \right) \Rightarrow x^{\mathsf{T}} P \, x \leq b. \qquad (4.17)$$

More generally, Lagrangian relaxation uses a similar approach to express a constraint in the objective function. It consists in adding to the objective function the inner product of the vector of constraints with a positive vector of the Euclidean space whose dimension is the number of constraints. Let us

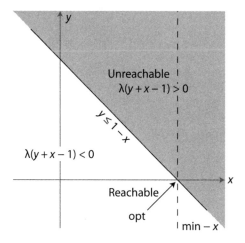

Figure 4.1. Example of a Lagrangian relaxation.

consider the following simple linear problem

$$\min_{x \in R} \quad \langle c, x \rangle$$
$$\text{s.t.} \quad ax \le b. \tag{4.18}$$

It is possible to express a second problem without constraints by introducing a nonnegative Lagrange multiplier $\lambda \in \mathbb{R}^+$:

$$\max_{\lambda \in \mathbb{R}^+} \min_{x \in \mathbb{R}} \langle c, x \rangle + \lambda(ax - b). \tag{4.19}$$

Since λ is positive, any x satisfying the constraint $ax \le b$ renders the term $\lambda(ax - b)$ negative. Trying to maximize the goal over variable λ, the optimum is obtained when $\lambda = 0$. Fixing x, any $\lambda' > \lambda$ will generate a solution $\langle c, x \rangle + \lambda'(ax - b) < \langle c, x \rangle + \lambda(ax - b)$. However, any x outside of the constraint will generate a positive term $\lambda(ax - b)$ that will be made arbitrarily large when trying to maximize it over $\lambda \ge 0$.

Example 4.11. *Figure 4.1 represents a simple Lagrangian relaxation in the linear case. The objective function is $-x$ when $y \le 1 - x$. The optimal solution is $(x, y) = (1, 0)$. When maximizing over λ in term $\lambda(y + x - 1)$, one obtains 0 when satisfying the constraint, and $+\infty$ otherwise.*

Similarly, a maximization problem can be reformulated as an inf sup when using Lagrangian relaxation to integrate a constraint in the objective.

$$\inf_{\substack{ax \leq b \\ x \in \mathbb{R}}} f(x) \leq \sup_{\lambda \in \mathbb{R}^+} \inf_{x \in \mathbb{R}} f(x) - \lambda(ax - b) \tag{4.20}$$

$$\sup_{\substack{ax \leq b \\ x \in \mathbb{R}}} f(x) \geq \inf_{\lambda \in \mathbb{R}^+} \sup_{x \in \mathbb{R}} f(x) - \lambda(ax - b) \tag{4.21}$$

In the case of equality constraint, any sign before the $\lambda(ax - b)$ term is valid.
These notions are easily extended to LMI as long as the constraint to integrate into the objective is conic convex and can be expressed as $A_i x - b_i \preceq 0$.

4.2.2.2   SOS extensions: SOS reinforcement and relaxation

The SOS reinforcement of polynomial optimization problems consists of restricting polynomial nonnegativity to being an element of $\Sigma[x]$. In the case of polynomial maximization problems, the SOS reinforcement boils down to computing an upper bound of the real optimal value. For example, let $p \in \mathbb{R}[x]$ and consider the unconstrained polynomial maximization problem $\sup \{p(x) \mid x \in \mathbb{R}^d\}$. Applying SOS reinforcement, we obtain:

$$\begin{aligned} \sup\{p(x) \mid x \in \mathbb{R}^d\} &= \inf\{\eta \mid \forall x, \eta - p(x) \geq 0\} \\ &\leq \inf\{\eta \mid \eta - p \in \Sigma[x]\}. \end{aligned} \tag{4.22}$$

Now, let $p, q \in \mathbb{R}[x]$ and consider the constrained polynomial maximization problem:

$$\sup\{p(x) \mid \forall x \in \mathbb{R}^d, q(x) \leq 0\}.$$

We can perform a Lagrangian relaxation but require λ to be a positive (SOS) polynomial instead of a positive scalar. Let $\lambda \in \Sigma[x]$, then:

$$\sup_{q(x) \leq 0, \ x \in \mathbb{R}^d} p(x) \leq \sup_{x \in \mathbb{R}^d} p(x) - \lambda(x) \cdot q(x).$$

Indeed, suppose $q(x) \leq 0$, then $-\lambda(x)q(x) \geq 0$ and $p(x) \leq p(x) - \lambda(x)q(x)$. Finally, taking the supremum over $\{x \in \mathbb{R}^d \mid q(x) \leq 0\}$ provides the above inequality. Since $\sup\{p(x) - \lambda(x) \cdot q(x) \mid x \in \mathbb{R}^d\}$ is an unconstrained polynomial maximization problem then we apply an SOS reinforcement (as in Eq. (4.22)) and we obtain:

$$\begin{aligned} \sup_{q(x) \leq 0, \ x \in \mathbb{R}^d} p(x) &\leq \sup_{x \in \mathbb{R}^d} p(x) - \lambda(x) \cdot q(x) \\ &\leq \inf\{\eta \mid \eta - p - \lambda q \in \Sigma[x]\}. \end{aligned} \tag{4.23}$$

Finally, note that this latter inequality is valid whatever $\lambda \in \Sigma[x]$ and so we can take the infimum over $\lambda \in \Sigma[x]$ which leads to:

$$\sup_{q(x) \leq 0, \ x \in \mathbb{R}^d} \quad p(x) \leq \inf_{\lambda \in \Sigma[x]} \sup_{x \in \mathbb{R}^d} \quad p(x) - \lambda(x) \cdot q(x)$$
$$\leq \inf_{\substack{\eta - p - \lambda q \in \Sigma[x] \\ \lambda \in \Sigma[x]}} \quad \eta \ . \tag{4.24}$$

4.2.3 Duality

A last useful manipulation of optimization problems relies on topological duality in Banach spaces (vector spaces with good topological structures). We give here an incomplete and informal overview of duality theory, since it enables the characterization of the dual problem. The interested reader can find more details in [46, §5.2].

Any vector space E over the field $\mathbb{R}$ can be associated with its dual vector space $E^\dagger$ defined as the set of real-valued linear functionals on E. That is the set of functions $\phi : E \to \mathbb{R}$. For any element of E we can associate an element of its dual space. This is characterized by the duality bracket $\langle \phi_x, x \rangle_{E^\dagger, E}$.

Thanks to the Riesz representation theorem, this element is unique in a Hilbert space and can be represented in the same space. Let's consider, for example, the finite dimensional Hilbert space $\mathbb{R}^n$, and an element $c \in \mathbb{R}^n$. The dual space is the set of linear functionals over $\mathbb{R}^n$, that is, the set of linear functions $\phi : \mathbb{R}^n \to \mathbb{R}$. Any such linear function can be defined by a scalar product. One can then build the linear functional associated to c: $\phi_c(x \in \mathbb{R}^n) = \langle x, c \rangle$ where $\langle \cdot, \cdot \rangle$ denotes the inner product of the Hilbert space E, in that case the scalar product of $\mathbb{R}^n$. Hilbert spaces are then auto-dual since a linear functional can be characterized by an element of the initial space.

Let us consider the general case of two Banach spaces E and F, $E^\dagger$ and $F^\dagger$ their topological dual, respectively, $K \subseteq E$ a convex cone (and $K^\dagger$ its dual), and the following optimizing problem:

$$\begin{aligned} \max \quad & \langle c, x \rangle_{E^\dagger, E} \\ \text{s.t.} \quad & Ax = b \text{ with } A : E \to F \\ & x \in K. \end{aligned} \tag{4.25}$$

In the following, let us denote this problem as the *primal problem*. This form is equivalent to the earlier version of Eq. (4.11) (cf. [46] for more explanation):

$$\begin{aligned} \max \quad & \langle c, x \rangle \\ \text{s.t.} \quad & A_i x - b_i \preceq_K 0 \text{ for } i \in [1, m] \\ & Ax = b. \end{aligned} \tag{4.26}$$

The constraint $Ax = b$ is equivalent to $Ax - b = 0_F$. Then one can introduce a Lagrangian multiplier $y \in F^\dagger$ to express the constraint. We have the duality bracket $\langle y, Ax - b \rangle_{F^\dagger, F}$. Using linearity of the linear form, one has $\langle y, Ax - b \rangle_{F^\dagger, F} = \langle y, Ax \rangle_{F^\dagger, F} - \langle y, b \rangle_{F^\dagger, F}$.

We can introduce the adjoint $A' : F^\dagger \to E^\dagger$ of A as the unique linear application such that $\langle y, Ax \rangle_{F^\dagger, F} = \langle A'y, x \rangle_{E^\dagger, E}$.

The constraint can then be expressed as $\langle A'y, x \rangle_{E^\dagger, E} - \langle y, b \rangle_{F^\dagger, F}$.

Going back to the initial problem, we have

$$\begin{array}{ll} & \forall y \in F^\dagger, \\ \max_x \langle c, x \rangle_{E^\dagger, E} & \leq \max_x \ \langle c, x \rangle_{E^\dagger, E} \\ \text{s.t} \left\{ \begin{array}{l} Ax = b \\ x \in K \end{array} \right. & \qquad + \langle A'y, x \rangle_{E^\dagger, E} \\ & \qquad - \langle y, b \rangle_{F^\dagger, F}. \end{array}$$

Since y is free in the left-hand part, one can build the following inequality:

$$\begin{array}{ll} \max_x \ \langle c, x \rangle_{E^\dagger, E} & \leq \min_y \max_x \ \langle c + A'y, x \rangle_{E^\dagger, E} \\ \text{s.t} \left\{ \begin{array}{l} Ax = b \\ x \in K \end{array} \right. & \qquad - \langle y, b \rangle_{F^\dagger, F}. \end{array}$$

The maximum with respect to x depends only on $\langle c + A'y, x \rangle$. Let us first define the dual cone of K as the restriction of the topological dual of E to positive linear forms on K:

$$K^\dagger = \{ f \in E^\dagger | \forall x \in K, \langle f, x \rangle \geq 0 \}.$$

Since $x \in K$, we have to make sure that $\langle c + A'y, x \rangle_{E^\dagger, E}$ will not diverge and corrupt the maximum of x when minimizing y. If we choose $c + A'y \in K^\dagger$, then the duality bracket is positive and will impact badly the maximum over x. Therefore, we have to choose $-c - A'y \in K^\dagger$.

We obtain the dual optimization problem:

$$\begin{array}{ll} \min_y & - \langle y, b \rangle_{F^\dagger, F} \\ \text{s.t.} & - c - A'y \in K^\dagger. \end{array} \tag{4.27}$$

While this description is general and will be used later in Sec. 5.6, a simpler version on Hilbert spaces will be used in chapter 6.

FEASIBILITY OF PRIMAL AND DUAL PROBLEMS

A last remark concerns the feasibility of the two primal and dual problems. Thanks to the construction of the dual problem and the use of Lagrangian relaxation, we have the following inequality:

$$\begin{array}{llll}
\max_x & \langle c, x \rangle_{E^\dagger, E} & \leq & \min_y & -\langle y, b \rangle_{F^\dagger, F} \\
\text{s.t.} & A : E \to F & & \text{s.t.} & -c - A'y \in K^\dagger \\
& Ax = b & & & \\
& x \in K. & & &
\end{array} \tag{4.28}$$

In the case where both optimization problems admit strict feasible solutions, we speak about primal and dual feasibility and in the case of convex constraints, the inequality of Eq. (4.28) becomes an equality, without any duality gap. In other words, solving any of the two problems gives the optimum solution. The conditions required are usually referred to as Slater's conditions.

In practice, one can easily obtain cases where one of the problems has an empty interior and is not strictly feasible. In that case the numerical solutions of both problems are not the same; we speak about a duality gap between these two solutions. We will come back to that notion in chapter 10.

Chapter Five

Invariant Synthesis via Convex Optimization

Postfixpoint Computation as Semialgebraic Constraints

5.1 INVARIANTS, LYAPUNOV FUNCTIONS, AND CONVEX OPTIMIZATION

This chapter focuses on the computation of invariant for a discrete dynamical system collecting semantics.

Invariants or collecting semantics properties are properties preserved along all executions of a system and verified in all reachable states. A subset of these invariants are defined as inductive. Inductive invariants are properties, or relationships between variables, that are inductively preserved by one transition of considered systems. Intuitively, it is not required to consider a reachable state and all (or part of) its past while arguing about the validity of the invariant, but only the single state. Applying the induction principle we obtain that any state satisfying the property is mapped to a next state preserving that same property.

For example, when one analyzes a geometric progression with a ratio r such that $|r| < 1$ then any invariant expressed as an interval can be easily proved: if $[a, b]$ contains the initial state and zero, any element of the progression will belong to $[a, b]$. Note that, here, we are only focused on the invariant but not interested in characterizing the decay or growth rate of the progression.

Discrete dynamical systems admit an infinite behavior, it is therefore of utmost importance to be able to characterize their reachable states, for example, proving the boundedness of such a set. In the control community *lingua* a system is said to be stable if, without any input, it converges to zero. This idea is captured by Lyapunov functions.

The current chapter proposes methods to compute dynamical systems' invariants based on Lyapunov function synthesis using convex optimization. In this section we introduce these notions. The following sections develop different encodings to compute these invariants for a wide variety of settings and solve different kinds of optimization problems.

5.1.1   Fixpoint characterization, invariant and inductive invariants

The motivation is to determine automatically if a given property holds for the analyzed program, or to compute precise bounds on reachable states. We are interested in numerical properties and more precisely in properties on the values taken by the d-uplet of the variables of the program.

According to the abstract interpretation framework outlined in Sec. 2.5, a semantics can be characterized by a set of elements; for collecting semantics that is the set of reachable states. Hence, in our point of view, as for the semantics characterization, a property is characterized by some set $P \subseteq \mathbb{R}^d$ of values satisfying the property.

Let us first recall the fixpoint characterization and instantiate it on our discrete dynamical system formalization.

5.1.1.1   *Collecting semantics as postfixpoint characterization*

In Sec. 2.5 we introduced the collecting semantics map in Eq. (2.16) and the fixpoint characterization of the collecting semantics in Eq. (2.17).

$$\begin{aligned} \wp(\Sigma) &\rightarrow \wp(\Sigma) \\ S &\mapsto I \cup f(S) \end{aligned} \tag{5.1}$$

$$\mathfrak{C} = lfp_\perp F = min_{X \in \wp(\Sigma)} \{F(X) \subseteq X\} \tag{5.2}$$

where f denotes the transition relation.

As a consequence any subset C of $\wp(\Sigma)$ verifying the condition $\{F(X) \subseteq X\}$ is a sound over-approximation of $\mathfrak{C}$ since all reachable states verify C: $\mathfrak{C} \subseteq C$.

5.1.1.2   *Collecting semantics of discrete dynamical systems*

Let us consider now a program of the forms presented in Sec. 4.1.1, 4.1.2, or 4.1.3. In the most general case, it is characterized by an initial set X^{Init}, and by a list of update functions f_i and associated conditions c_i. In the following we assume that we are given a set representation X^i of each condition c_i. We recall that X^i are assumed to form a partition of Σ, i.e., for each element a unique update function is applicable.

Let C be the set satisfying the previous equation, over-approximating reachable states $\mathfrak{C}$. With F a piecewise discrete dynamical system, we have the following constraints on P:

$$\begin{aligned} &\{F(C) \subseteq C\} \\ &= \{X^{Init} \cup f(C) \subseteq C\} \\ &= \left\{ C \,\middle|\, \begin{array}{l} X^{Init} \subseteq C \\ \text{for } i \in \mathcal{I},\, f_i(C \cap X^i) \subseteq C \end{array} \right\}. \end{aligned} \tag{5.3}$$

This equation can be further simplified in the case of a single update function, i.e., no disjunction nor conditions c_i, X^i.

$$\left\{ C \;\middle|\; \begin{array}{l} X^{Init} \subseteq C \\ f(C) \subseteq C \end{array} \right\} \tag{5.4}$$

5.1.2 Lyapunov functions

In 1890, Alexander Lyapunov published his well-known result stating that the differential equation $\frac{d}{dt}x = Ax(t)$ is stable if and only if there exists a positive definite matrix P such that $A^\intercal P + PA \preceq 0$. Here both A and P are square matrices of $\mathbb{R}^{n \times n}$ and P is positive definite $P \succ 0$, i.e., $\forall x \in \mathbb{R}^n, x^\intercal Px \geq 0$. Later this was formulated in a discrete-time setting over discrete linear systems:

$$x_{k+1} = Ax_k \text{ with } A \in \mathbb{R}^{n \times n}$$

as

$$\exists P \in\in \mathbb{R}^{n \times n}, \text{ s.t. } \left\{ \begin{array}{l} \exists P \succ 0 \\ A^\intercal PA - P \preceq 0. \end{array} \right. \tag{5.5}$$

In both cases, P is the measure of energy of the system: the Lyapunov function $x \mapsto x^\intercal Px$. When *measuring* the energy of the image state Ax, we obtain $(Ax)^\intercal P(Ax) = x^\intercal A^\intercal PAx$.

Since P is positive definite, $\forall x \in \mathbb{R}^n, x^\intercal Px > 0$, and P denotes a norm over states, while, thanks to the second constraint, its sublevel sets are inductive over states: $\forall x \in \mathbb{R}^n, x^\intercal Px \geq x^\intercal A^\intercal PAx$. The inequality $A^\intercal PA - P \preceq 0$ encodes a kind of energy dissipation along trajectories. When the energy reaches 0, the state of the system is near 0. In this original setting, the considered system is closed, i.e., it does not admit input. In the case of linear systems with bounded inputs, one can rely on the same argument not to motivate asymptotic stability but to argue that the system will not diverge and remains within some bounds.

Simpler arguments do exist for the specific case of linear systems, e.g., one can compute the eigenvalue of the matrix and check that the linear map A is contracting. However, this notion of Lyapunov function seems more extensible and was widely developed in the control community.

Let us consider numerical systems with $\Sigma = \mathbb{R}^d$. More formally, a Lyapunov function $V : \mathbb{R}^d \to \mathbb{R}^+$ for a discrete-time system is a positive real-valued function over system states that should satisfy:

- Null at origin, positive elsewhere

$$\left\{ \begin{array}{l} V(0) = 0 \\ \forall x \in \mathbb{R}^d \backslash \{0\}, V(x) > 0 \wedge \lim_{\|x\| \to \infty} V(x) = \infty. \end{array} \right. \tag{5.6}$$

- Decreasing along trajectories

$$\forall x \in \mathbb{R}^d, V \circ f(x) - V(x) \leq 0. \tag{5.7}$$

Depending on the strictness of the $\leq$ operator, the Lyapunov function guaranties asymptotic stability and exponential convergence, or just boundedness of states.

It is shown, for example in [80], that exhibiting such a function proves the Lyapunov stability of the system, meaning that its state variables will remain bounded through time. Equation (5.7) expresses the fact that the function $k \mapsto V(x_k)$ decreases, which, combined with (5.6), shows that the state variables remain in the bounded sublevel set $\{x \in \mathbb{R}^n | V(x) \leq V(x_0)\}$ at all instances $k \in \mathbb{N}$.

5.1.3 Lyapunov functions as problem specific abstractions: semialgebraic template abstractions

We saw that Lyapunov functions characterize inductive sublevel sets for the considered discrete dynamical system semantics. Therefore, instead of approximating reachable states in the abstract interpretation framework using predefined numerical abstractions, such as intervals, octagons, or convex polyhedra, we rather propose to rely on the Lyapunov function as the main mean of abstraction. This is a template abstraction [90, 99].

A template is a real-valued function $t : \Sigma \to \mathbb{R}$.

Example 5.1. *A template is then a function over those state variables. For example, it can characterize the norm 2 of a state:*

$$t_1(s) = ||s||_2 = \sqrt{(\Sigma_{v \in V} v^2)}$$

or just focus on the value of a single variable $x \in V$

$$
\begin{aligned}
t_2(s) &= s(x) && \text{when } s \text{ is a map or} \\
t_2(s) &= x_i && \text{when } s \text{ is a vector.}
\end{aligned}
$$

For a given template t, a sublevel set abstraction can be defined by a given levelset λ:

$$\{s | t(s) \leq \lambda\}.$$

In the case of multiple templates $t_1, t_2, \ldots, t_n$ and associated bounds $\lambda_1, \lambda_2, \ldots, \lambda_n$, the interpretation of this abstract representation is the intersection of sublevel sets. In the case of polynomial template functions t_i, this is a basic semialgebraic set.

$$\bigcap_i \{s | t_i(s) \leq \lambda_i\}$$

5.1.4   Synthesis of templates using convex optimization

In the early definition of Lyapunov functions, with quadratic properties and linear systems, the conditions defining the Lyapunov function P were characterized as a Linear Matrix Inequality (LMI):

$$\begin{cases} \exists P \succeq 0 \\ A^\mathsf{T} P A - P \preceq 0. \end{cases} \tag{5.8}$$

With the development of interior point algorithms [24] and convex optimization [46], the numerical resolution of these optimization problems becomes feasible in reasonable time.

The approach is to guide the search for inductive invariants as Lyapunov-like constraints expressed as convex optimization problems.

Once a Lyapunov function is synthesized, as a norm of a state, it can be used as a template abstraction and denote a relevant abstraction of reachable states. Depending on the encoding of the constraints, the results of the optimization step could either be a bound template, e.g., $t(x) \le 1$, or just a relevant unbounded template t.

In that second case, the template t has to be bounded by other means; for example, using classical Kleene iterations, or even using randomly large values. Thanks to the inductiveness property of the template with respect to the system semantics, any bound λ such that

$$t(f(x)) \le \lambda$$

characterizes a sound postfixpoint (invariant):

$$\{x \,|\, t(x) \le \lambda\}.$$

5.2   QUADRATIC INVARIANTS

5.2.1   Linear systems

As mentioned above, in the simplest case of a linear system the conditions over a quadratic Lyapunov function P are given by the LMI of Equation (5.5).

$$\exists P \in \mathbb{R}^{n \times n}, \text{ s.t.} \begin{cases} \exists P \succeq 0 \\ A^\mathsf{T} P A - P \preceq 0 \end{cases} \tag{5.9}$$

One can directly solve this LMI and obtain a valid quadratic template, relevant for the considered system. However, while inductive over system semantics, a sublevel set property characterized by such Lyapunov function P may not be the most precise:

$$\mathfrak{C} \subset \left\{x \in \mathbb{R}^d \,|\, x^\mathsf{T} P x \le \lambda\right\}.$$

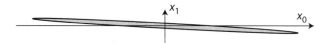

Figure 5.1. "Flat" ellipsoids can yield very large bounds on some variables.

In order to synthesize a more precise invariant, one can further constrain the LMI.

5.2.1.1   Minimizing condition number

Graphically, the condition number of a positive definite matrix expresses a notion similar to that addressed by eccentricity for ellipses in dimension 2. It measures how "close" to a circle (or its higher dimension equivalent) the resulting ellipsoid will be. Multiples of the identity matrix, which all represent a circle, have a condition number of 1. Thus one idea of constraint we can impose on P is to have its condition number as close to 1 as possible. A rationale for this is that "flat" ellipsoids, i.e., having a large condition number, can yield a very bad bound on one of the variables, as illustrated in Figure 5.1.

This is done [23] by minimizing a new variable, r, in the following matrix inequality

$$I \preceq P \preceq rI.$$

Indeed, if a point x is in the ellipsoid P, then $x^\mathsf{T} P x \leq 1$ which implies $x^\mathsf{T} I x \leq 1$, i.e., x is in the sphere of radius 1. Thus, the ellipsoid P is included in the sphere of radius 1. Similarly, P contains the sphere of radius $r^{-\frac{1}{2}}$. This way, P is sandwiched between these two spheres and making their radius as close as possible will make P as "round" as possible, as depicted in Figure 5.2.

This constraint, along with the others (Lyapunov equation, symmetry, and positive definiteness of P), can be expressed as an LMI, which is solved using the semidefinite programming techniques mentioned in Section 4.2.1.2:

$$\begin{aligned}
\text{minimize} \quad & r \\
\text{subject to} \quad & A^\mathsf{T} P A - P \prec 0 \\
& I \preceq P \preceq rI \\
& P^\mathsf{T} = P.
\end{aligned} \tag{5.10}$$

$$A^\mathsf{T} P A - r P \preceq 0. \tag{5.11}$$

Example 5.2.   *With the following matrix A of the running example*

$$A := \begin{bmatrix} 0.9379 & -0.0381 & -0.0414 \\ -0.0404 & 0.968 & -0.0179 \\ 0.0142 & -0.0197 & 0.9823 \end{bmatrix}$$

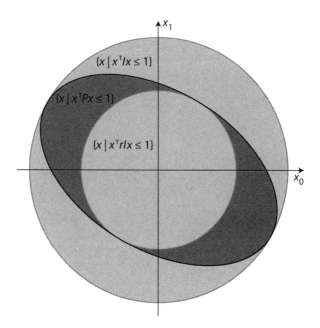

Figure 5.2. Making the ellipsoid P as "round" as possible by sandwiching it between spheres of radius $r^{-\frac{1}{2}}$ and 1: $I \preceq P \preceq rI$ and minimizing r.

a semidefinite solver simply returns $r = 1$ and the identity matrix

$$P = \begin{bmatrix} 1 & 0 & 0 \\ 0 & 1 & 0 \\ 0 & 0 & 1 \end{bmatrix}.$$

5.2.1.2 Preserving the shape

Another approach [20] is to minimize $r \in (0,1)$ in the following inequality. Intuitively, this corresponds to finding the shape of ellipsoid that gets "preserved" the best when the update $x_{k+1} = Ax_k$ is applied, as depicted in Figure 5.3. r can be seen as the minimum contraction achieved by this update in the norm defined by P, hence the name *decay rate* given to this value by control theorists. This is the choice implicitly made in [47] for a particular case of matrices A of order 2.

With this technique, however, the presence of a quadratic term rP in the equation prevents the use of usual LMI solving tools "as is." To overcome this, the following property enables the choice of an approach where the value for r is refined by dichotomy. Only a few steps are then required to obtain a good approximation of the optimal value.

Property 5.3. *If Equation (5.11) admits as solution a positive definite matrix P for a given r, then it is also the case for any $r' \geq r$.*

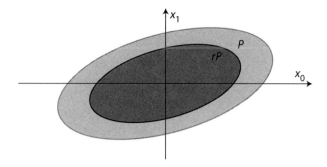

Figure 5.3. Choice of the ellipsoid whose shape is the best preserved.

Example 5.4. *With the following matrix A of the running example:*

$$A := \begin{bmatrix} 0.9379 & -0.0381 & -0.0414 \\ -0.0404 & 0.968 & -0.0179 \\ 0.0142 & -0.0197 & 0.9823 \end{bmatrix},$$

looking for a small $r \in (0,1)$, the first value tested is $r = 0.5$, i.e., a solution to the following semidefinite program is looked for:

$$\begin{aligned} \text{minimize} \quad & 0 \\ \text{subject to} \quad & A^{\mathsf{T}} P A - 0.5 P \preceq 0 \\ & P \succ 0 \\ & P^{\mathsf{T}} = P. \end{aligned}$$

Since there is no solution, r is now looked for in interval $(0.5, 1)$. $r = 0.75$ is tested, without more success, then $r = 0.875$, $r = 0.9375$, $r = 0.98675$, and $r = 0.984375$ are still unsuccessful. Finally, $r = 0.9921875$ yields the following solution (all figures being rounded to four digits):

$$P = \begin{bmatrix} 239.1338 & 37.5557 & 77.9203 \\ 37.5557 & 226.3640 & 65.8287 \\ 77.9203 & 65.8287 & 325.1628 \end{bmatrix}.$$

Stopping here leaves $r \in (0.984375, 0.9921875)$ and the above matrix P as the solution for $r = 0.9921875$.

5.2.2   Consider linear systems with inputs

Most system trajectories are not purely characterized by their initial state: they have inputs.

$$x_{k+1} = A x_k + B u_k, \|u_k\|_\infty \leq 1. \tag{5.12}$$

In the case of unbounded input the system is guaranteed to diverge. We are therefore interested in showing that, when the input values at bounded $\|u_k\|_\infty \leq 1$ (i.e., $\max_k u_k \leq 1$), then the system still has a bounded behavior. This constraint over u_k is reasonable: most inputs come from sensors which themselves have physical limits. We can also choose the bound 1 without loss of generality since one can always alter the matrix B to account for different bounds.

Considering the inputs requires a slight reinforcement of Equation (5.5) into

$$A^\mathsf{T} P A - P \prec 0. \tag{5.13}$$

We can still guarantee that the state variables of (5.12) will remain in a sub-level set $\{x \in \mathbb{R}^n \mid x^\mathsf{T} P x \leq \lambda\}$ (for some $\lambda > 0$), which is an ellipsoid in this case.

5.2.2.1 Quadratic invariant for bounded-input linear systems

The two previous methods were based only on A, completely abstracting B away, which could lead to rather coarse abstractions. We try here to take both A and B into account by finding the ellipsoid P included in the smallest possible sphere which is stable, i.e., such that

$$\forall x, \forall u, \|u\|_\infty \leq 1 \wedge x^\mathsf{T} P x \leq 1,$$
$$(Ax + Bu)^\mathsf{T} P (Ax + Bu) \leq 1.$$

This is illustrated in Figure 5.4. The previous condition can be rewritten as

$$\forall x, \forall u, \left(\bigwedge_{i=0}^{p-1} (e_i^\mathsf{T} u)^2 \leq 1 \right) \wedge x^\mathsf{T} P x \leq 1$$
$$\Rightarrow (Ax + Bu)^\mathsf{T} P (Ax + Bu) \leq 1$$

where e_i is the i-th vector of the canonical basis (i.e., with all coefficients equal to 0 except the i-th one which is 1). This amounts to

$$\forall x, \forall u, \left(\bigwedge_{i=0}^{p-1} \begin{bmatrix} x \\ u \end{bmatrix}^\mathsf{T} \begin{bmatrix} 0 & 0 \\ 0 & E^{i,i} \end{bmatrix} \begin{bmatrix} x \\ u \end{bmatrix} \leq 1 \right)$$
$$\wedge \begin{bmatrix} x \\ u \end{bmatrix}^\mathsf{T} \begin{bmatrix} P & 0 \\ 0 & 0 \end{bmatrix} \begin{bmatrix} x \\ u \end{bmatrix} \leq 1$$
$$\Rightarrow \begin{bmatrix} x \\ u \end{bmatrix}^\mathsf{T} \begin{bmatrix} A^\mathsf{T} P A & A^\mathsf{T} P B \\ B^\mathsf{T} P A & B^\mathsf{T} P B \end{bmatrix} \begin{bmatrix} x \\ u \end{bmatrix} \leq 1$$

where $E^{i,j}$ is the matrix with 0 everywhere except the coefficient at line i, column j which is 1. Using the S-Procedure (Theorem 4.10), this holds when there are

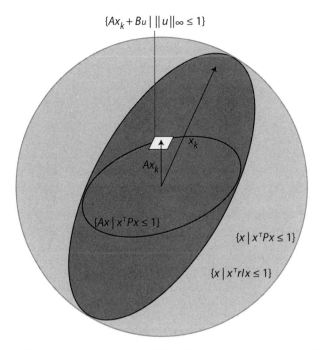

Figure 5.4. Looking for an invariant ellipsoid included in the smallest possible sphere by maximizing r.

τ and $\lambda_0, \ldots, \lambda_{p-1}$ all nonnegatives such that

$$
\begin{bmatrix} -A^\mathsf{T} PA & -A^\mathsf{T} PB & 0 \\ -B^\mathsf{T} PA & -B^\mathsf{T} PB & 0 \\ 0 & 0 & 1 \end{bmatrix} - \tau \begin{bmatrix} -P & 0 & 0 \\ 0 & 0 & 0 \\ 0 & 0 & 1 \end{bmatrix}
$$
$$
- \sum_{i=0}^{p-1} \lambda_i \begin{bmatrix} 0 & 0 & 0 \\ 0 & -E^{i,i} & 0 \\ 0 & 0 & 1 \end{bmatrix} \succeq 0.
$$

(5.14)

As in Section 5.2.1.2, this is not an LMI since τ and P are both variables. And again, there is a $\tau_{min} \in (0,1)$ such that this inequality admits as solution a positive definite matrix P if and only if $\tau \in (\tau_{min}, 1)$. This value τ_{min} can, by the way, be approximated thanks to the exact same procedure. Similarly to what was done in Section 5.2.1.1, P is forced to be contained in the smallest possible sphere by maximizing r in the additional constraint

$$
P \succeq rI. \tag{5.15}
$$

The function f is then defined as the function mapping $\tau \in (\tau_{min}, 1)$ to the optimal value of the following semidefinite program.

$$\text{maximize} \quad r$$

$$\text{subject to} \quad (5.14)$$

$$(5.15)$$

$$P^{\mathsf{T}} = P \tag{5.16}$$

$$\bigwedge_{i=0}^{p-1} \lambda_i \geq 0$$

Figure 5.5. Quadratic template for bounded-input linear systems.

This function f can then be evaluated for a given input τ simply by solving the above semidefinite program. f seems concave which could enable a smart optimization procedure. However, in practice, it is enough to just sample f for some equally spaced values in the interval $(\tau_{min}, 1)$ and just keep the matrix P obtained for the value enabling the greatest r.

Example 5.5. *With the following matrices A and B of the running example:*

$$A := \begin{bmatrix} 0.9379 & -0.0381 & -0.0414 \\ -0.0404 & 0.968 & -0.0179 \\ 0.0142 & -0.0197 & 0.9823 \end{bmatrix} B := \begin{bmatrix} 0.0237 \\ 0.0143 \\ 0.0077 \end{bmatrix},$$

according to Example 5.4, $\tau_{min} = 0.9921875$.
    Then, f is evaluated on a few points between τ_{min} and 1 (rounded figures):

τ	$f(\tau)$	τ	$f(\tau)$
0.9928	1.6064	0.9967	0.7440
0.9935	1.4653	0.9974	0.5970
0.9941	1.3231	0.9980	0.4490
0.9948	1.1798	0.9987	0.3002
0.9954	1.0355	0.9993	0.1505
0.9961	0.8902		

and the one giving the best value ($\tau = 0.9928$) is kept with the corresponding

$$P = \begin{bmatrix} 12.6465 & -14.1109 & -10.5402 \\ -14.1109 & 25.6819 & 3.06577 \\ -10.5402 & 3.06577 & 29.5981 \end{bmatrix}.$$

5.2.2.2   Optimize template for a given variable

If a tighter bound is required on one of the variables, the identity matrix I in inequality (5.15) can be replaced by a diagonal matrix with larger coefficients

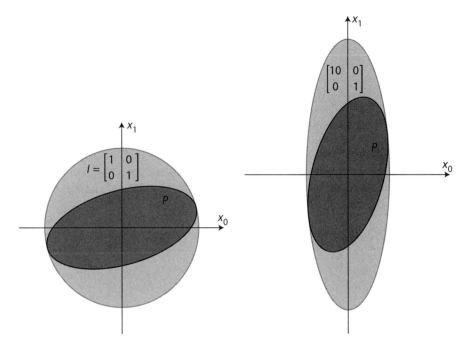

Figure 5.6. *(Left)*: Constraining ellipsoid P to lie in a sphere *(Right)*: Constraining ellipsoid P to lie in an ellipsoid flatter in a given direction.

for variables of interest. For instance, to get a smaller bound on the first variable x_0, the matrix I can be replaced by $\begin{bmatrix} 10 & 0 \\ 0 & I \end{bmatrix}$.

This intuitively corresponds to minimize the radius of an ellipsoid containing P flatter on the dimension of interest instead of a sphere. This is illustrated in Figures 5.8a and 5.8b.

Example 5.6. *With the following matrices A and B of the running example:*

$$A := \begin{bmatrix} 0.9379 & -0.0381 & -0.0414 \\ -0.0404 & 0.968 & -0.0179 \\ 0.0142 & -0.0197 & 0.9823 \end{bmatrix} B := \begin{bmatrix} 0.0237 \\ 0.0143 \\ 0.0077 \end{bmatrix},$$

expressing a higher interest in the first variable as exposed above gives

$$P = \begin{bmatrix} 12.6465 & -14.1109 & -10.5402 \\ -14.1109 & 25.6819 & 3.06577 \\ -10.5402 & 3.06577 & 29.5981 \end{bmatrix}.$$

5.3 PIECEWISE QUADRATIC INVARIANTS

5.3.1 Piecewise affine systems

While strong results do exist for pure linear systems, most of them vanish in the presence of nonlinearity such as switches between linear dynamics. As we saw in the previous section, stable linear systems were guaranteed to admit a quadratic Lyapunov function and therefore a quadratic invariant. In a switched linear system, this property is undecidable [37, Theorem 2]. The proposed methods are therefore meant to be understood as heuristics, trying to synthesize a meaningful invariant for such systems.

As presented in Sec. 4.1.2 describing switched linear systems, these systems are composed by a set of linear updates A_i associated to conditions c_i. A common practice in control is to look for a common Lyapunov function: a quadratic Lyapunov function characterized by a positive definite matrix P such that

$$A_1^\mathsf{T} P A_1 - A_1 \preceq 0$$
$$A_2^\mathsf{T} P A_2 - A_2 \preceq 0$$
$$\cdots$$
$$A_n^\mathsf{T} P A_n - A_n \preceq 0.$$

This common Lyapunov function decreases along trajectories regardless of the active cell (see Sec. 4.1.2.1). Note that, in this encoding, all information about the condition satisfied in each cell is ignored.

The main difficulty in the switched case is related to the change of dynamics: we must decrease whenever a transition from one cell to another is fired. Moreover, we only require the norm induced by the quadratic Lyapunov function P to be local, i.e., positive only where the law is used.

Therefore, the main goal is to synthesize a Lyapunov function $V(x, u)$ and an associated bound α characterizing the invariant of reachable states as a sublevel set S_α, such that

$$\forall i \in \mathcal{I}, \forall (x, u) \in X^i, V(x, u) \leq \alpha \tag{5.17}$$

$$\forall i, j \in \mathcal{I}, \forall (x, u) \in X^i, \forall (x', u') \in X^j, \text{ s.t.}$$
$$x' = A^i x + B^i u + b^i, V(x, u) \geq V(x', u'). \tag{5.18}$$

5.3.2 Encode conditions and switches as quadratic constraints

In Equations (5.17) and (5.18), the inequalities on V are local on cells. In (5.17), the function has to decrease only on feasible transitions from cell X^i to cell X^j.

In order to encode the problem as a set of linear matrix inequalities, we need to express conditions associated to each cell in suitable form. For SDP, encoding constraints requires us to be able to express cell membership or feasible transitions as quadratic constraints.

5.3.2.1   Quadratization of cells

We recall that for us local means that it is true on a cell and thus true on a poly-hedron. Using the homogeneous version of a cell, we can define local positive-ness on a polyhedral cone. Let Q be a $d \times d$ symmetric matrix and M be an $n \times d$ matrix. Local positivity in this case means that $My \geq 0 \implies y^\mathsf{T} Q y \geq 0$. The problem will be to write the local positivity as a constraint without impli-cation. The problem is not new (e.g., the survey paper [34]). [10] proves that local positivity is equivalent, when M has a full row rank, to $Q - M^\mathsf{T} C M \succeq 0$ where C is a copositive matrix, i.e., $x^\mathsf{T} C x \geq 0$ if $x \geq 0$. First, in general (when the rank of M is not necessarily equal to its number of rows), note that if $Q - M^\mathsf{T} C M \succeq 0$ for some copositive matrix C then Q satisfies $My \geq 0 \implies y^\mathsf{T} Q y \geq 0$. Secondly, every matrix C with nonnegative entries is coposi-tive. Since copositivity seems to be as difficult as local positivity to handle, we will restrict copositive matrices to be matrices with nonnegative entries. The idea is instead of using cells as polyhedral cones, we use a quadratization of cells by introducing nonnegative entries and we will define the quadratization of a cell X^i by:

$$\overline{X^i} = \left\{ \begin{pmatrix} x \\ u \end{pmatrix} \in \mathbb{R}^{d+m} \;\middle|\; \begin{pmatrix} 1 \\ x \\ u \end{pmatrix}^\mathsf{T} E^{i\mathsf{T}} W^i E^i \begin{pmatrix} 1 \\ x \\ u \end{pmatrix} \geq 0 \right\} \tag{5.19}$$

where W^i is a $(1+n_i) \times (1+n_i)$ symmetric matrix with nonnegative entries and $E^i = \begin{pmatrix} E_s^i \\ E_w^i \end{pmatrix}$ with $E_s^i = \begin{pmatrix} 1 & 0_{1 \times (d+m)} \\ c_s^i & -T_s^i \end{pmatrix}$ and $E_w^i = \begin{pmatrix} c_w^i & -T_w^i \end{pmatrix}$. Recall that n_i is the number of rows of T^i. The matrix E^i is thus of the size $n_i + 1 \times (1 + d + m)$. The goal of adding the row $(1, 0_{1 \times (d+m)})$ is to avoid adding the opposite of a vector of X^i in $\overline{X^i}$. Indeed without this latter vector $\overline{X^i}$ would be symmetric. We illustrate this fact in Example 5.7. Note that during the optimization process, matrices W^i will be decision variables.

Example 5.7 (Homogenization).   *Let us take the polyhedra* $X = \{x \in \mathbb{R} \mid x \leq 1\}$. *Using our notations, we have* $X = \{x \mid M(1 \ x)^\mathsf{T} \geq 0\}$ *with* $M = (1 \ -1)$. *Let us consider two cases, the first one without adding the row and the second one using it.*

  Without any modification, the quadratization of X *relative to a nonnega-tive real* W *is* $X' = \{x \mid (1 \ x) M^\mathsf{T} W M (1 \ x)^\mathsf{T} \geq 0\}$. *But* $(1 \ x) M^\mathsf{T} W M (1 \ x)^\mathsf{T} = W(1 \ x)(1 \ -1)^\mathsf{T}(1 \ -1)(1 \ x)^\mathsf{T} = 2W(1 - x)^2$. *Hence,* $X' = \mathbb{R}$ *for all nonnegative real* W.

  Now let us take $E = \begin{pmatrix} 1 & 0 \\ 1 & -1 \end{pmatrix}$. *The quadratization as defined by Equation* (5.19) *relative to a* 2×2 *symmetric matrix* W *with nonnegative coefficients is* $\overline{X} = \{x \mid (1 \ x) E^\mathsf{T} W E (1 \ x)^\mathsf{T} \geq 0\}$. *We have:*

$$(1\ x) \begin{pmatrix} 1 & 1 \\ 0 & -1 \end{pmatrix} \begin{pmatrix} w_1 & w_3 \\ w_3 & w_2 \end{pmatrix} \begin{pmatrix} 1 & 0 \\ 1 & -1 \end{pmatrix} (1\ x)^\mathsf{T}$$

$$= w_1 + 2w_3(1 - x) + w_2(1 - x)^2.$$

Then any matrix W such that $w_2 = w_1 = 0$ and $w_3 > 0$ implies that $\overline{X} = X$.

Now we introduce an example of the quadratization of the cell X^1 for our running example.

Example 5.8. *Let us consider the running example and the cell X^1. We recall that X^1 is characterized by the matrices and vectors:*

$$\begin{cases} T_s^1 = \begin{pmatrix} -9 & 7 & 6 \\ -4 & 8 & -8 \end{pmatrix} \\ c_s^1 = (5\ 4)^\mathsf{T} \end{cases},$$

$$\begin{cases} T_w^1 = \begin{pmatrix} 0 & 0 & 1 \\ 0 & 0 & -1 \end{pmatrix} \\ c_w^1 = (3\ 3)^\mathsf{T} \end{cases}$$

$$and\ E^1 = \begin{pmatrix} 1 & 0 & 0 & 0 \\ 5 & 9 & -7 & -6 \\ 4 & 4 & -8 & 8 \\ 3 & 0 & 0 & -1 \\ 3 & 0 & 0 & 1 \end{pmatrix}.$$

As suggested we have added the row $(1, 0_{1 \times 3})$. Take, for example, the matrix:

$$W^1 = \begin{pmatrix} 63.0218 & 0.0163 & 0.0217 & 12.1557 & 8.8835 \\ 0.0163 & 0.0000 & 0.0000 & 0.0267 & 0.0031 \\ 0.0217 & 0.0000 & 0.0000 & 0.0094 & 0.0061 \\ 12.1557 & 0.0267 & 0.0094 & 4.2011 & 59.5733 \\ 8.8835 & 0.0031 & 0.0061 & 59.5733 & 3.0416 \end{pmatrix}.$$

We have

$$\overline{X^1} = \{(x, y, u) \mid (1, x, y, u) E^1 W^1 E^1 (1, x, y, u)^\mathsf{T} \geq 0\}$$
$$\supseteq\ X^1$$

Local positivity of quadratic forms will also be used when a transition from a cell to an other is fired. For the moment, we are interested in the set of (x, u) such that $(x, u) \in X^i$ and whose image is in X^j, and we denote by X^{ij} the set:

$$\left\{ \begin{pmatrix} x \\ u \end{pmatrix} \in \mathbb{R}^{d+m} \middle| \begin{array}{c} \begin{pmatrix} x \\ u \end{pmatrix} \in X^i \text{ and} \\ (A^i x + B^i u + b^i, u) \in X^j \end{array} \right\}$$

for all pairs $i, j \in \mathcal{I}$. Note that in [35], the authors take into account all pairs (i, j) such that there exists a state x_k at time k in X^i and the image of x_k that is x_{k+1} is in X^j. We will discuss in Subsection 5.3.2.2 the computation or a reduction to possible switches using linear programming as suggested in [51]. To construct a quadratization of X^{ij}, we use the same approach as before by introducing a $(1 + n_i + n_j) \times (1 + n_i + n_j)$ symmetric matrix U^{ij} with nonnegative entries to get a set $\overline{X^{ij}}$ defined as:

$$\overline{X^{ij}} = \left\{ \begin{pmatrix} x \\ u \end{pmatrix} \in \mathbb{R}^{d+m} \middle| \begin{pmatrix} 1 \\ x \\ u \end{pmatrix}^{\mathsf{T}} E^{ij\,\mathsf{T}} U^{ij} E^{ij} \begin{pmatrix} 1 \\ x \\ u \end{pmatrix} \geq 0 \right\} \tag{5.20}$$

where $E^{ij} = \begin{pmatrix} E_s^{ij} \\ E_w^{ij} \end{pmatrix}$ with

$$E_s^{ij} = \begin{pmatrix} 1 & 0_{1 \times (d+m)} \\ c_s^i & -T_s^i \\ c_s^j - T_s^j \begin{pmatrix} b^i \\ 0 \end{pmatrix} & -T_s^j \begin{pmatrix} A^i & B^i \\ 0_{d \times m} & \mathrm{Id}_{m \times m} \end{pmatrix} \end{pmatrix}$$

and

$$E_w^{ij} = \begin{pmatrix} c_w^i & -T_w^i \\ c_w^j - T_w^j \begin{pmatrix} b^i \\ 0 \end{pmatrix} & -T_w^j \begin{pmatrix} A^i & B^i \\ 0_{d \times m} & \mathrm{Id}_{m \times m} \end{pmatrix} \end{pmatrix}. \tag{5.21}$$

5.3.2.2 Switching cells

We have to manage another constraint which comes from the cell switches. After applying the available law in cell X^i, we have to specify the reachable cells, i.e., the cells X^j such that there exists (x, u) satisfying:

$$(x, u) \in X^i \text{ and } (A^i x + B^i u + b^i, u) \in X^j.$$

We say that a switch from i to j is fireable iff:

$$
\left\{ (x,u) \in \mathbb{R}^{d+m} \middle| \begin{array}{l} T_s^i(x,u)^{\mathsf{T}} \ll c_s^i \\[4pt] T_s^j(A^i x + B^i u + b^i, u)^{\mathsf{T}} \ll c_s^j \\[4pt] T_w^i(x,u)^{\mathsf{T}} \le c_w^i \\[4pt] T_w^j(A^i x + B^i u + b^i, u)^{\mathsf{T}} \le c_w^j \end{array} \right\} \tag{5.22}
$$

$$
\neq \emptyset.
$$

We will denote by $i \to j$ if the switch from i to j is fireable. Recall that the symbol $<$ means that we can deal with both strict inequalities and inequalities. Problem (5.22) is a linear programming feasibility problem with both strict and weak inequalities. However, we only check whether the system is solvable and we can detect infeasibility by using Motzkin transposition theorem [1]. Motzkin's theorem is an alternative type theorem, that is, we oppose two linear systems such that exactly one of the two is feasible. To describe the alternative system, we have to separate strict and weak inequalities and use the matrices E_s^{ij} and E_w^{ij} defined in Equation (5.21). Problem (5.22) is equivalent to check whether the set $\{y = (z, x, u) \in \mathbb{R}^{1+d+m} \mid E_w^{ij} y \ge 0, \; E_s^{ij} y \gg 0\}$ is empty or not. To detect feasibility we test the infeasibility of the alternative system defined as:

$$
\left\{ \begin{array}{l} (E_s^{ij})^{\mathsf{T}} p^s + (E_w^{ij})^{\mathsf{T}} p = 0 \\[10pt] \sum_{k \in \mathbb{I}} p_k^s = 1 \\[10pt] p_k^s \ge 0, \; \forall k \in \mathbb{I} \\[10pt] p_i \ge 0, \; \forall i \notin \mathbb{I}. \end{array} \right. \tag{5.23}
$$

From Motzkin's transposition theorem [1], we get the following proposition.

Proposition 5.9. *Problem (5.22) is feasible iff Problem (5.23) is not.*

However, reasoning directly on the matrices can allow unfireable switches. For certain initial conditions, for all $k \in \mathbb{N}$, the condition $(x_k, u_k) \in X^i$ and $(A^i x_k + B^i u + b^i, u) \in X^j$ does not hold whereas Problem (5.22) is feasible. To avoid it, we must know all the possible trajectories of the system (which we want to compute) and remove all inactivated switches. A sound way to under-approximate unfireable transitions is to identify unsatisfiable sets of linear constraints.

Example 5.10. *We continue to detail our running example. More precisely, we consider the possible switches. We take, for example, the cell X^2. To switch from cell X^2 to cell X^1 is possible if the following system of linear inequalities has a solution:*

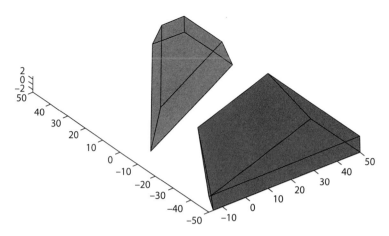

Figure 5.7. The truncated representation of X^2 in light gray and the preimage of X^1 by the law inside X^2 in dark gray.

$$
\begin{aligned}
-9x + 7y + 6u &< 5 \\
-0.8532x + 2.5748y - 10.4460 &< -68 \\
-3.3662x + 2.1732y - 1.1084u &< -58 \\
4x - 8y + 8u &\leq -4 \\
u &\leq 3 \\
-u &\leq 3.
\end{aligned}
\tag{5.24}
$$

The first two consist in constraining the image of (x, y, u) to belong to X^1 and the last four constraints correspond to the definition of X^2. The representation of these two sets (X^2 and the preimage of X^1 by the law defined in X^2) is given in Figure 5.7. We see in Figure 5.7 that the system of inequalities defined in Equation (5.24) seems to not have solutions. We check that using Equation (5.23) and Proposition 5.9. The matrices E_s^{ij} and E_w^{ij} of Equation (5.23) are in this example:

$$
E_s^{21} = \begin{pmatrix} 5 & 9 & -7 & -6 \\ -68 & 0.8532 & -2.5748 & 10.446 \\ -58 & 3.3662 & -2.1732 & 1.1084 \end{pmatrix}
$$

$$
\text{and } E_w^{21} = \begin{pmatrix} -4 & -4 & 8 & -8 \\ 3 & 0 & 0 & -1 \\ 3 & 0 & 0 & 1 \end{pmatrix}.
$$

We thus solved the linear program defined in Equation (5.23) (with MATLAB and Linprog) and we found $p = (0.8735, 0.0983, 0.0282)^\mathsf{T}$ and $q = (0.3325, 14.2500, 7.8461)^\mathsf{T}$. This means that the alternative system is feasible and consequently the *initial is not from Proposition 5.9. Finally, the transition from X^2 to X^1 is not possible.*

5.3.3   Local invariants with coupling conditions

As in the linear case, we are relying here on SDP solver and LMI encoding; the unknowns of the optimization problems have to be at most quadratic.

5.3.3.1   *Piecewise quadratic Lyapunov function*

The Lyapunov function V is piecewise defined, relying on the partition of cells provided by the analyzed piecewise affine system. This V is defined as:

$$
\begin{aligned}
V(x, u) &= V^i(x, u), \text{ if } \begin{pmatrix} x \\ u \end{pmatrix} \in X^i \\
&= \begin{pmatrix} x \\ u \end{pmatrix}^{\mathsf{T}} P^i \begin{pmatrix} x \\ u \end{pmatrix} + 2{q^i}^{\mathsf{T}} \begin{pmatrix} x \\ u \end{pmatrix}, \text{ if } \begin{pmatrix} x \\ u \end{pmatrix} \in X^i.
\end{aligned}
$$

The function V^i is thus a local function only defined on X^i.

A sublevel set S_α of V of level $\alpha \in \mathbb{R}$ is represented as:

$$
\begin{aligned}
S_\alpha &= \bigcup_{i \in \mathcal{I}} S_{i,\alpha} \\
&= \bigcup_{i \in \mathcal{I}} \left\{ \begin{pmatrix} x \\ u \end{pmatrix} \in X^i \mid \begin{pmatrix} x \\ u \end{pmatrix}^{\mathsf{T}} P^i \begin{pmatrix} x \\ u \end{pmatrix} + 2{q^i}^{\mathsf{T}} x \leq \alpha \right\} \\
&= \bigcup_{i \in \mathcal{I}} \left\{ \begin{pmatrix} x \\ u \end{pmatrix} \in X^i \mid \begin{pmatrix} 1 \\ x \\ u \end{pmatrix}^{\mathsf{T}} \begin{pmatrix} -\alpha & {q^i}^{\mathsf{T}} \\ q^i & P^i \end{pmatrix} \begin{pmatrix} 1 \\ x \\ u \end{pmatrix} \leq 0 \right\}.
\end{aligned}
$$

The set $S_{i,\alpha}$ is thus the local sublevel set of V^i associated to the level α.

So we are looking at a family of pairs of a matrix and a vector $\{(P^i, q^i)\}_{i \in \mathcal{I}}$ and a real $\alpha \in \mathbb{R}$ such that S_α is invariant by the piecewise affine system. To obtain invariance property, we have to constrain S_α to contain initial conditions of the system. Finally, to prove that the reachable set is bounded, we have to constrain S_α to be bounded.

Before deriving the semidefinite constraints, let us first state a useful result in Proposition 5.11. This result, which is a special case of the S-Procedure (4.10), allows us to encode implications into semidefinite constraints in a safe way. The implication must involve quadratic inequalities on both sides.

Proposition 5.11.   *Let A, B, C be $d \times d$ matrices. Then, $C + A + B \succeq 0$ implies that the implication $(y^{\mathsf{T}} A y \leq 0 \wedge y^{\mathsf{T}} B y \leq 0) \implies y^{\mathsf{T}} C y \geq 0$ holds.*

5.3.3.2 *Writing invariance as semidefinite constraints*

We assume that $(x, u) \in X^i \cap S_{i,\alpha}$ (this index i is unique). Invariance means that if we apply the available law to (x, u) and suppose that the image of (x, u) belongs to some cell X^j (notation $i \to j$), then the image of (x, u) belongs to $S_{j,\alpha}$. Note that $(x, u) \in X^i$ and its image is supposed to be in X^j then $(x, u) \in X^{ij}$. Let $(i, j) \in \mathcal{I}^2$ such that $i \to j$, invariance translated in inequalities and implication gives:

$$\begin{pmatrix} x \\ u \end{pmatrix} \in X^{ij} \wedge \begin{pmatrix} x \\ u \end{pmatrix} \in S_{i,\alpha}$$

$$\Longrightarrow \begin{pmatrix} A^i x + B^i u + b^i \\ u \end{pmatrix} \in S_{j,\alpha}. \tag{5.25}$$

We can use the relaxation of Subsection 5.3.2.1 as a representation of cells and use matrix variables W^i and U^{ij} to encode their quadratization. We get for $(i, j) \in \mathcal{I}^2$ such that $i \to j$:

$$\begin{pmatrix} 1 \\ x \\ u \end{pmatrix}^\mathsf{T} E^{ij\,\mathsf{T}} U^{ij} E^{ij} \begin{pmatrix} 1 \\ x \\ u \end{pmatrix} \geq 0$$

$$\wedge \quad \begin{pmatrix} 1 \\ x \\ u \end{pmatrix}^\mathsf{T} \begin{pmatrix} -\alpha & q^{i\,\mathsf{T}} \\ q^i & P^i \end{pmatrix} \begin{pmatrix} 1 \\ x \\ u \end{pmatrix} \leq 0 \tag{5.26}$$

$$\Longrightarrow \begin{pmatrix} 1 \\ x \\ u \end{pmatrix}^\mathsf{T} \left(F^{i\,\mathsf{T}} \begin{pmatrix} -\alpha & q^{j\,\mathsf{T}} \\ q^j & P^j \end{pmatrix} F^i \right) \begin{pmatrix} 1 \\ x \\ u \end{pmatrix} \leq 0$$

where E^{ij} is the matrix defined in Equation (5.20) and F^i is defined in Equation (4.4).

Finally, we obtain a stronger condition by considering a semidefinite constraint such as Equation (5.27). Proposition 5.11 proves that if $(P^i, P^j, q^i, q^j, U^{ij})$ is a solution of Equation (5.27) then $(P^i, P^j, q^i, q^j, U^{ij})$ satisfies Equation (5.26). For $(i, j) \in \mathcal{I}^2$ such that $i \to j$:

$$-F^{i\,\mathsf{T}} \begin{pmatrix} 0 & q^{j\,\mathsf{T}} \\ q^j & P^j \end{pmatrix} F^i + \begin{pmatrix} 0 & q^{i\,\mathsf{T}} \\ q^i & P^i \end{pmatrix} - E^{ij\,\mathsf{T}} U^{ij} E^{ij} \succeq 0. \tag{5.27}$$

Note that the symbol $-\alpha$ is canceled during the computation.

5.3.4   Initialization and boundedness

5.3.4.1   Integrating initial conditions

To complete invariance property, invariant set must contain initial conditions. Suppose that initial condition is a polyhedron $X^0 = \{(x, u) \in \mathbb{R}^{d+m} \mid T_w^0(x, u) \leq c_w^0, \; T_s^0(x, u) \ll c_s^0\}$. We must have $X^0 \subseteq S_\alpha$. But X^0 is contained in the union of X^i. Hence, X^0 is the union over $i \in \mathcal{I}$ of the sets $X^0 \cap X^i$. If, for all $i \in \mathcal{I}$, the set $X^0 \cap X^i$ is contained in $S_{i,\alpha}$ then $X^0 \subseteq S_\alpha$. We can use the same method as before to express that all sets $S_{i,\alpha}$ such that $X^0 \cap X^i \neq \emptyset$ must contain $X^0 \cap X^i$. In terms of implications, it can be rewritten as for all $i \in \mathcal{I}$ such that $X^0 \cap X^i \neq \emptyset$:

$$(x, u) \in X^0 \cap X^i \implies (x, u) P^i (x, u)^\mathsf{T} + 2(x, u)q^i \leq \alpha. \tag{5.28}$$

Since $X^0 \cap X^i$ is a polyhedra, it admits some quadratization, that is, $\overline{X^0 \cap X^i} = \{(x, u) \in \mathbb{R}^{d+m} \mid (1, x, u) E^{0i\mathsf{T}} Z^i E^{0i} (1, x, u)^\mathsf{T} \geq 0\}$ where $E^{0i} = \begin{pmatrix} E_s^{0i} \\ E_w^{0i} \end{pmatrix}$ with:

$$E_w^{0i} = \begin{pmatrix} c_w^0 & -T_w^0 \\ c_w^i & -T_w^i \end{pmatrix} \text{ and } E_s^{0i} = \begin{pmatrix} 1 & 0_{1\times(d+m)} \\ c_s^0 & -T_s^0 \\ c_s^i & -T_s^i \end{pmatrix}$$

and Z^i is some symmetric matrix whose coefficients are nonnegative.

For all $i \in \mathcal{I}$ such that $X^0 \cap X^i \neq \emptyset$, we obtain a stronger notion by introducing semidefinite constraints:

$$-\begin{pmatrix} -\alpha & q^{i\mathsf{T}} \\ q^i & P^i \end{pmatrix} - E^{0i\mathsf{T}} Z^i E^{0i} \succeq 0. \tag{5.29}$$

Proposition 5.11 proves that if (P^i, q^i, Z^i) is a solution of Equation (5.29) then (P^i, q^i, Z^i) satisfies Equation (5.28).

Note, since $X^0 \cap X^i$ is a polyhedron then its emptiness can be decided by checking the feasibility of the linear problem (5.30) and by using the same argument as Proposition 5.9.

$$\begin{cases} (E_s^{0i})^\mathsf{T} p^s + (E_w^{0i})^\mathsf{T} p = 0 \\ \sum_{k \in \mathbb{I}} p_k^s = 1 \\ p_k^s \geq 0, \; \forall k \in \mathbb{I} \\ p_i \geq 0, \; \forall i \notin \mathbb{I} \end{cases} \tag{5.30}$$

Linear program (5.30) is feasible iff $X^0 \cap X^i = \emptyset$.

input : Piecewice affine system defined by $T^i_{s,w}, c^i_{s,w}, A^i, B^i, b^i, \forall i \in \mathcal{I}$
local : $E^i, E^{ij}, E^{0i}, \forall i, j \in \mathcal{I}$
output: $\alpha, \beta, P^i, q^i, Z^i, W^i, U^{ij}, \forall i, j \in \mathcal{I}$

1 Compute quadratization of cells E^i using Equation (5.19), $\forall i \in \mathcal{I}$;
2 Over-approximate feasible switches: compute possible switches $L \in \mathcal{I}^2$ using Equation (5.22);
3 Compute quadratization of switches E^{ij} using Equation (5.20), $\forall i, j \in L$;
4 Compute quadratization of initialization E^{0i} using Equation (5.29), $\forall i \in \mathcal{I}$;
5 Solve the SDP problem of Equation (5.34)
6 Invariants:
7 $\bigcup_{i \in \mathcal{I}} \left\{ (x, u) P^i_{opt}(x, u)^\mathsf{T} + 2(x, u) q^i_{opt} \leq \alpha_{opt} \, \big| (x, u) \in X^i \right\}$
8 $\|(x, u)\| \leq \beta_{opt}$

Figure 5.8. Algorithm to compute piecewise quadratic invariant for piecewise affine dynamical systems.

5.3.4.2 *Writing boundedness as semidefinite constraints*

The sublevel S_α is bounded if and only if for all $i \in \mathcal{I}$, the sublevel $S_{i,\alpha}$ is bounded. The boundedness constraint in terms of implications is, for all $i \in \mathcal{I}$, there exists $\beta \geq 0$:

$$(x, u) \in X^i \wedge \begin{pmatrix} x \\ u \end{pmatrix} \in S_{i,\alpha} \implies \|(x, u)\|_2^2 \leq \beta \qquad (5.31)$$

where $\| \cdot \|_2$ denotes the Euclidean norm of $\mathbb{R}^{d+m}$.

As invariance, we use the quadratization of X^i and the definition of $S_{i,\alpha}$. We use the fact that $\|(x, u)\|_2^2 = \begin{pmatrix} x \\ u \end{pmatrix}^\mathsf{T} \mathrm{Id}_{(d+m) \times (d+m)} \begin{pmatrix} x \\ u \end{pmatrix}$ and we get for all $i \in \mathcal{I}$:

$$\begin{pmatrix} 1 \\ x \\ u \end{pmatrix}^\mathsf{T} E^{i\mathsf{T}} W^i E^i \begin{pmatrix} 1 \\ x \\ u \end{pmatrix} \geq 0 \wedge$$

$$\begin{pmatrix} 1 \\ x \\ u \end{pmatrix}^\mathsf{T} \begin{pmatrix} -\alpha & q^{i\mathsf{T}} \\ q^i & P^i \end{pmatrix} \begin{pmatrix} 1 \\ x \\ u \end{pmatrix} \leq 0 \implies \qquad (5.32)$$

$$\begin{pmatrix} 1 \\ x \\ u \end{pmatrix}^\mathsf{T} \begin{pmatrix} -\beta & 0_{1 \times (d+m)} \\ 0_{(d+m) \times 1} & \mathrm{Id}_{(d+m) \times (d+m)} \end{pmatrix} \begin{pmatrix} 1 \\ x \\ u \end{pmatrix} \leq 0$$

where E^i is defined in Equation (5.19).

Finally, as invariance we obtain a stronger condition by considering a semi-definite constraint such as Equation (5.33). Proposition 5.11 proves that (P^i, q^i, W^i) is a solution of Equation (5.33); the (P^i, q^i, W^i) satisfies Equation (5.32).

For all $i \in \mathcal{I}$:
$$-E^{i^\mathsf{T}} W^i E^i + \begin{pmatrix} -\alpha & q^{i^\mathsf{T}} \\ q^i & P^i \end{pmatrix}$$

$$+ \begin{pmatrix} \beta & 0_{1 \times (d+m)} \\ 0_{(d+m) \times 1} & -\operatorname{Id}_{(d+m) \times (d+m)} \end{pmatrix} \succeq 0. \tag{5.33}$$

5.3.5 Overall method

The algorithm in Figure 5.8 summarizes the method.

5.3.6 Example

The method applied to our piecewise affine system defined in Sec. 4.1.2 computes the following values:
$$\alpha_{opt} = 242.0155$$
$$\beta_{opt} = 2173.8501.$$

This means that $\|(x, y, u)\|_2^2 = x^2 + y^2 + u^2 \leq \beta_{opt}$. We can conclude, for example, that the values taken by the variables x are between $[-46.6154, 46.6154]$.

Note that this specific example does not admit a common Lyapunov function.

The value α_{opt} gives the level of the invariant sublevel of our piecewise quadratic Lyapunov function where the local quadratic functions are characterized by the following matrices and vectors:

minimize $\alpha + \beta$

$st. \forall i \in \mathcal{I}, (i, j) \in L,$

$$-F^{i^\mathsf{T}} \begin{pmatrix} 0 & q^{j^\mathsf{T}} \\ q^j & P^j \end{pmatrix} F^i + \begin{pmatrix} 0 & q^{i^\mathsf{T}} \\ q^i & P^i \end{pmatrix} - E^{ij^\mathsf{T}} U^{ij} E^{ij} \succeq 0 \ .$$

$$- \begin{pmatrix} -\alpha & q^{i^\mathsf{T}} \\ q^i & P^i \end{pmatrix} - E^{0i^\mathsf{T}} Z^i E^{0i} \succeq 0 \tag{5.34}$$

$$-E^{i^\mathsf{T}} W^i E^i + \begin{pmatrix} -\alpha & q^{i^\mathsf{T}} \\ q^i & P^i \end{pmatrix}$$

$$+ \begin{pmatrix} \beta & 0_{1 \times (d+m)} \\ 0_{(d+m) \times 1} & -\operatorname{Id}_{(d+m) \times (d+m)} \end{pmatrix} \succeq 0$$

Figure 5.9. Summary of generated SDP problem for piecewise affine discrete systems.

$$P^1 = \begin{pmatrix} 1.0181 & -0.0040 & -1.1332 \\ -0.0040 & 1.0268 & -0.5340 \\ -1.1332 & -0.5340 & -13.7623 \end{pmatrix}$$

$$q^1 = (0.1252, 1.3836, -29.6791)^{\mathsf{T}}$$

$$P^2 = \begin{pmatrix} 9.1540 & -7.0159 & -2.6659 \\ -7.0159 & 9.5054 & -2.4016 \\ -2.6659 & -2.4016 & -8.9741 \end{pmatrix}$$

$$q^2 = (-21.3830, -44.6291, 114.2984)^{\mathsf{T}}$$

$$P^3 = \begin{pmatrix} 1.1555 & -0.3599 & -2.6224 \\ -0.3599 & 2.4558 & -2.8236 \\ -2.6224 & -2.8236 & -2.3852 \end{pmatrix}$$

$$q^3 = (-5.3138, 6.7894, -40.5537)^{\mathsf{T}}$$

$$P^4 = \begin{pmatrix} 3.7314 & -3.4179 & -3.1427 \\ -3.4179 & 6.1955 & 0.9499 \\ -3.1427 & 0.9499 & -10.6767 \end{pmatrix}$$

$$q^4 = (28.5011, -73.5421, 48.2153)^{\mathsf{T}}.$$

Finally, for conciseness reasons, we do not provide here the matrix certificates W^i for each cell X^i, nor the matrices U^{ij} encoding quadratization matrices of polyhedron Xij. These matrices are computed by the analysis but do not provide useful information with respect to bounds.

5.4   K-INDUCTIVE QUADRATIC INVARIANTS

5.4.1   K-induction principle

The principle behind all computed invariants up to now was the inductiveness of computed Lyapunov function $V(x)$ with respect to the system transition function f.

However, as mentioned in Sec. 5.1.1, a property could be valid, i.e., an invariant, without being directly inductive. In SMT-based model-checking, a trade-off to prove the validity of a property for a given transition system $(\Sigma, I \subseteq \Sigma, T \in \Sigma^2)$ is to search for a k-induction proof [36, 102] instead of a 1-induction one.

In k-induction, the base step addresses the property verification on all traces of length up to k, rooted in an initial state, while the inductive step intends to show that any trace suffix of length k validating the property preserves it in the $k+1$-th step.

Definition 5.12 (*k*-induction). *Let* (Σ, I, T) *be a transition system over states* Σ *with initial states* $I \subseteq \Sigma$ *and transition relation* $T \subseteq \Sigma \times \Sigma$. *A safety property* $Prop \subseteq \Sigma$ *is said to be k-inductive with respect to the transition system iff*

- *for all system traces of length less than k, all reachable states verify Prop*

$$\forall j \leq k \in \mathbb{N}, \forall x_0, \ldots, x_j \in \Sigma,$$

$$x_0 \in I \; \wedge \bigwedge_{i \in [0, j-1]} (x_i, x_{i+1}) \in T \tag{5.35}$$

$$\implies x_j \in Prop$$

- *for all system subtraces of length k satisfying Prop then the next state satisfies Prop as well*

$$\forall x_0, \ldots, x_k \in \Sigma,$$

$$\bigwedge_{i \in [0, k-1]} x_i \in Prop \wedge (x_i, x_{i+1}) \in T \tag{5.36}$$

$$\implies x_k \in Prop.$$

In our fixpoint characterization, this amounts to substituting

$$\{F(C) \subseteq C\} = \{X^{Init} \cup f(C) \subseteq C\}$$

with

$$\begin{aligned} &\quad \left\{ F^k(C) \subseteq C \right\} \\ &= \left\{ X^{Init} \cup \bigcup_{1 \leq i \leq k} f^i(Init) \cup f^k(C) \subseteq C \right\} \\ &= \left\{ C \left| \begin{array}{l} X^{Init} \subseteq C \\ f(X^{Init}) \subseteq C \\ f^2(X^{Init}) \subseteq C \\ \ldots f^k(X^{Init}) \subseteq C \\ f(C) \subseteq C \end{array} \right. \right\}. \end{aligned}$$

5.4.2 *k*-inductive Lyapunov function

We recall that we consider a piecewise system composed of cells X^i indexed by a set $\mathcal{I}$ of partition labels, such that $\Sigma = \bigcup_{i \in \mathcal{I}} X^i$, and which transition relation is piecewise defined with transitions T^i. The *k*-inductive property $Prop$ denotes here a boundedness property represented by a sublevel set S^i_α of a

Lyapunov function V. Then, a k-induction proof amounts to finding this function V such that:

$$\forall j < k \in \mathbb{N}, \forall i_0, \dots, i_j \in \mathcal{I}, \forall x_0, \dots, x_j \in \Sigma,$$

$$x_0 \in (I \cap X^0) \wedge \bigwedge_{i \in [0, j-1]} x_i \in X^i \wedge (x_i, x_{i+1}) \in T^i \tag{5.37}$$

$$\implies x_j \in S_\alpha$$

$$\forall i_0, \dots, i_k \in \mathcal{I}, \forall x_0, \dots, x_k \in \Sigma,$$

$$\bigwedge_{i \in [0, k-1]} x_i \in (X^i \cap S_\alpha) \wedge (x_i, x_{i+1}) \in T^i \tag{5.38}$$

$$\implies x_k \in S_\alpha.$$

Let I^* be the set of finite words of the letters in I, and I_k^* its restriction to words of length exactly k. In the following, we denote by $|w|$ the length of word w, by $a \cdot b$ the concatenation of the words a and b into ab, and by $tl(w)$ the tail of a nonempty word w, i.e., w without its first letter. For example, $tl(i \cdot w) = w$.

Following the Lee and Dullerud approach [71, 72, 103], we reinforce the Equations (5.37)–(5.38) and search for a quadratic Lyapunov function V^w for each nonempty sequence of switches $w = i_0 \cdot \dots \cdot i_{k-1} \in \mathcal{I}_k^*$:

$$V^w \begin{pmatrix} x \\ u \end{pmatrix} = \begin{pmatrix} x \\ u \end{pmatrix}^t P^w \begin{pmatrix} x \\ u \end{pmatrix}.$$

Let $S_{w \cdot i, \alpha}$ be the local quadratic sublevel set associated to the nonempty path $w \cdot i$ and the level α:

$$S_{w \cdot i, \alpha} =$$

$$\left\{ \begin{pmatrix} x \\ u \end{pmatrix} \in X^i \; \middle| \; \begin{array}{l} V^{w \cdot i}(x, u) \leq \alpha \wedge \exists \begin{pmatrix} x' \\ u' \end{pmatrix} \text{ s.t.} \\[2mm] \left(\begin{pmatrix} x' \\ u' \end{pmatrix}, \begin{pmatrix} x \\ u \end{pmatrix} \right) \in T^j \wedge \\[2mm] \begin{pmatrix} x' \\ u' \end{pmatrix} \in S_{w, \alpha} \\[2mm] \text{when } w = w' \cdot j \end{array} \right\}.$$

Let us consider a nonempty finite path w; the sublevel $S_{w, \alpha}$ denotes that the $|w|$ predecessors of $\begin{pmatrix} x \\ u \end{pmatrix}$ belong to the sublevel associated to the path prefixes, e.g.,

$$S_{123,\alpha} =$$

$$\left\{ \begin{pmatrix} x \\ u \end{pmatrix} \in X^3 \;\middle|\; \begin{array}{c} V^{123}(x,u) \leq \alpha \wedge \exists \begin{pmatrix} x' \\ u' \end{pmatrix} \\[2mm] T\begin{pmatrix} x' \\ u' \end{pmatrix} = \begin{pmatrix} x \\ u \end{pmatrix} \wedge \begin{pmatrix} x' \\ u' \end{pmatrix} \in S_{12,\alpha} \end{array} \right\}$$

$$S_{12,\alpha} =$$

$$\left\{ \begin{pmatrix} x \\ u \end{pmatrix} \in X^2 \;\middle|\; \begin{array}{c} V^{12}(x,u) \leq \alpha \wedge \exists \begin{pmatrix} x' \\ u' \end{pmatrix} \\[2mm] T\begin{pmatrix} x' \\ u' \end{pmatrix} = \begin{pmatrix} x \\ u \end{pmatrix} \wedge \begin{pmatrix} x' \\ u' \end{pmatrix} \in S_{1,\alpha} \end{array} \right\}$$

$$S_{1,\alpha} = \left\{ \begin{pmatrix} x \\ u \end{pmatrix} \in X^1 \;\middle|\; V^1(x,u) \leq \alpha \right\}.$$

The equations can be rephrased as:

$$\forall\, w \in \mathcal{I}_1^*, \forall x \in \Sigma, x \in \mathcal{I} \cap X^i \implies x \in S_{w,\alpha} \tag{5.39}$$

$$\forall\, 1 \leq j < k, \forall w \cdot i \in \mathcal{I}_j^*, \forall x, y \in \Sigma,$$
$$(x,y) \in T^i \wedge x \in S_{w \cdot i,\alpha} \implies y \in S_{w \cdot i \cdot j,\alpha} \tag{5.40}$$

$$\forall\, w \cdot i \in \mathcal{I}_k^*, \forall x, y \in \Sigma,$$
$$(x,y) \in T^i \wedge x \in S_{w \cdot i,\alpha} \implies y \in S_{tl(w \cdot i) \cdot j,\alpha}. \tag{5.41}$$

Proposition 5.13. *Any solution $\{P^w | \forall 1 \leq j \leq k, w \in \mathcal{I}_j^*\}$ of Equations (5.39)–(5.41) satisfies (5.37)–(5.38) with S_α^i defined as*

$$S_\alpha^i = \left\{ \begin{pmatrix} x \\ u \end{pmatrix} \;\middle|\; \max_{w \cdot i \in \mathcal{I}^*} V^{w \cdot i} \begin{pmatrix} x \\ u \end{pmatrix} \leq \alpha \right\}.$$

We now adapt the semidefinite constraints of the previous section to satisfy the k-inductive based constraints. While it is possible to target directly the synthesis of a k-inductive piecewise quadratic sublevel set, the approach typically starts from $k = 1$ and increases to $k + 1$ in case of failure to find a minimal k-inductive piecewise quadratic invariant.

5.4.3   Associate quadratic invariants to path suffixes

We adapt the previous method to express properties over bounded sequence of past transitions.

5.4.3.1   Characterizing the graph of possible switches–enumerating the paths

As a first step, we compute the set of possible paths of given length up to k. First, a graph $\mathcal{G} = (I, Init, Switches)$ denoting possible switches between cells $i \in \mathcal{I}$ is computed using the approach presented in Sec. 5.3.2.2.

$Init = \{i \in \mathcal{I} | X^0 \cap X^i \neq \emptyset\}$ denotes the subset of cells $i \in \mathcal{I}$ that verify the initial conditions. This characterizes a set of polyhedral constraints for which vacuity is computed using the method presented in Sec. 5.3.2.1.

We then enumerate the possible paths in the graph using classical graph algorithms. Let $Paths^k$ be such a set of paths of length up to k.

Example 5.14.   *The Figure 5.10 presents the possible transitions as over-approximated by the method presented in Sec. 5.3.2.2. Depending on the target length the following paths are generated:*

length	
1	1,2,3,4
2	11, 12, 13, 14, 22, 24, 31, 33, 34, 41, 42, 43, 44
3	111, 112, 113, 114, 122, 124, 131, 133, 134, 141, 142, 143, 144, 222, 224, 241, 242, 243, 244, 311, 312, 313, 314, 331, 333, 334, 341, 342, 343, 344, 411, 412, 413, 414, 422, 424, 431, 433, 434, 441, 442, 443, 444
4	...

5.4.3.2   Integrating initial conditions

The initial condition only applies for the quadratic sublevel associated to initial cells. Let $Init$ be the set of cells admitting initial elements, as defined in the graph construction.

By construction of the set of paths $Paths^k$, it contains the single letter words denoting initial cells $\{i \mid i \in Init\} \subseteq Paths^k$. The set of initial constraints only apply for the one letter word satisfying the initial condition:

$$(x, u) \in X^0 \cap X^i \implies (x, u) P^i (x, u)^\mathsf{T} + 2(x, u) q^i \leq \alpha. \qquad (5.42)$$

We can rely on the same stronger encoding as a semidefinite constraint, using the quadratization of the condition $X^0 \cap X^i$ as the matrix E^{0i}:

$$-\begin{pmatrix} -\alpha & q^{i\mathsf{T}} \\ q^i & P^i \end{pmatrix} - E^{0i\mathsf{T}} Z^i E^{0i} \succeq 0. \qquad (5.43)$$

Note that, independently of the value of k, a system with n cells is para-metrized by at most n Z^i variables.

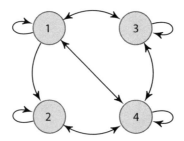

Figure 5.10. Switch graph of the running example.

5.4.3.3 Expressing transitions in initial and inductive cases as semidefinite constraints

Equations (5.40) and (5.41) denoting a transition X^{ij} from cell X^i to cell X^j can be defined as:

$$
\begin{pmatrix} x \\ u \end{pmatrix} \in X^{ij} \wedge \begin{pmatrix} x \\ u \end{pmatrix} \in S_{w \cdot i, \alpha}
$$
$$
\implies \begin{pmatrix} A^i x + B^i u + b^i \\ u \end{pmatrix} \in S_{w \cdot i \cdot j, \alpha}
\tag{5.44}
$$

$$
\begin{pmatrix} x \\ u \end{pmatrix} \in X^{ij} \wedge \begin{pmatrix} x \\ u \end{pmatrix} \in S_{w \cdot i, \alpha} \wedge |w \cdot i| = k
$$
$$
\implies \begin{pmatrix} A^i x + B^i u + b^i \\ u \end{pmatrix} \in S_{tl(w \cdot i) \cdot j, \alpha}.
\tag{5.45}
$$

As before, these constraints are first relaxed with the use of quadratization of cell transitions E^{ij}, and then expressed as semidefinite constraints using Prop. 5.11.

when $|i \cdot w| = k$:

$$
-F^{i\mathsf{T}} \begin{pmatrix} 0 & q^{tl(w \cdot i) \cdot j\,\mathsf{T}} \\ q^{tl(w \cdot i) \cdot j} & P^{tl(w \cdot i) \cdot j} \end{pmatrix} F^i
$$
$$
+ \begin{pmatrix} 0 & q^{w \cdot i\,\mathsf{T}} \\ q^{w \cdot i} & P^{w \cdot i} \end{pmatrix} - E^{ij\,\mathsf{T}} U^{w \cdot i, j} E^{ij} \succeq 0.
\tag{5.46}
$$

when $|i \cdot w| < k$:

$$
-F^{i\mathsf{T}} \begin{pmatrix} 0 & q^{w \cdot i \cdot j\,\mathsf{T}} \\ q^{w \cdot i \cdot j} & P^{w \cdot i \cdot j} \end{pmatrix} F^i
$$
$$
+ \begin{pmatrix} 0 & q^{w \cdot i\,\mathsf{T}} \\ q^{w \cdot i} & P^{w \cdot i} \end{pmatrix} - E^{ij\,\mathsf{T}} U^{w \cdot i, j} E^{ij} \succeq 0.
\tag{5.47}
$$

input : Piecewise affine system defined by $T^i_{s,w}, c^i_{s,w}, A^i, B^i, b^i, \forall i \in \mathcal{I}$
local : $E^i, E^{ij}, E^{0i}, \forall i, j \in \mathcal{I}$
output: $\alpha, \beta, P^{w\cdot i}, q^{w\cdot i}, Z^i, W^{w\cdot i}, U^{w\cdot i,j}, \forall i, j \in \mathcal{I}, w \in Paths^k$

1 Compute quadratization of cells E^i using Equation (5.19), $\forall i \in \mathcal{I}$;
2 Over-approximate feasible switches: compute possible switches $L \in \mathcal{I}^2$ using
   Equation (5.22);
3 Compute $Paths^k$ list of paths of length $\leq k$;
4 Compute quadratization of switches E^{ij} using Equation (5.20), $\forall i, j \in L$;
5 Compute quadratization of initialization E^{0i} using Equation (5.29), $\forall i \in \mathcal{I}$;
6 Solve the SDP problem of Equation (5.50)
7 Invariants:

8 $\displaystyle\bigcup_{w\cdot i \in Paths^k} \left\{ \begin{array}{l} (x,u)P^{w^c doti}_{opt}(x,u)^\mathsf{T} + 2(x,u)q^{w\cdot i}_{opt} \leq \alpha_{opt} \\ \text{when } (x,u) \in X^i \end{array} \right\}$

9 $\|(x,u)\| \leq \beta_{opt}$

Figure 5.11. Algorithm to compute piecewise k-inductive quadratic invariant for piecewise affine dynamical systems.

Note that we have $|Paths^k|$ variables q^w, P^w and $|Paths^k| \times |I|$ variables $U^{w,j}$.

5.4.3.4 *Expressing boundedness*

The boundedness constraint expressed as a semidefinite constraint is straightforward. We require that all path-associated quadratic sublevels are bounded by the same scalar β.

For all $w \cdot i \in Paths^k$, there exists $\beta \geq 0$:

$$(x,u) \in X^i \wedge \begin{pmatrix} x \\ u \end{pmatrix} \in S_{w\cdot i, \alpha} \implies \|(x,u)\|^2_2 \leq \beta. \tag{5.48}$$

The associated semidefinite constraints are:

$$-E^{i\mathsf{T}} W^{w\cdot i} E^i + \begin{pmatrix} -\alpha & q^{w\cdot i\mathsf{T}} \\ q^{w\cdot i} & P^{w\cdot i} \end{pmatrix}$$

$$+ \begin{pmatrix} \beta & 0_{1\times(d+m)} \\ 0_{(d+m)\times 1} & -\operatorname{Id}_{(d+m)\times(d+m)} \end{pmatrix} \succeq 0. \tag{5.49}$$

We have here $|Paths^k|$ variables W^w.

minimize $\alpha + \beta$

$st. \forall i \in \mathcal{I}, (i,j) \in L,$

$\forall w \in Paths^k, \text{ s.t. } |i \cdot w| = k$

$$-F^{i\mathsf{T}} \begin{pmatrix} 0 & q^{tl(w \cdot i) \cdot j\mathsf{T}} \\ q^{tl(w \cdot i) \cdot j} & P^{tl(w \cdot i) \cdot j} \end{pmatrix} F^i$$

$$+ \begin{pmatrix} 0 & q^{w \cdot i\mathsf{T}} \\ q^{w \cdot i} & P^{w \cdot i} \end{pmatrix} - E^{ij\mathsf{T}} U^{w \cdot i, j} E^{ij} \succeq 0 .$$

$\forall w \in Paths^k, \text{ s.t. } |i \cdot w| < k$

$$-F^{i\mathsf{T}} \begin{pmatrix} 0 & q^{w \cdot i \cdot j\mathsf{T}} \\ q^{w \cdot i \cdot j} & P^{w \cdot i \cdot j} \end{pmatrix} F^i$$

$$+ \begin{pmatrix} 0 & q^{w \cdot i\mathsf{T}} \\ q^{w \cdot i} & P^{w \cdot i} \end{pmatrix} - E^{ij\mathsf{T}} U^{w \cdot i, j} E^{ij} \succeq 0 .$$

$$-E^{i\mathsf{T}} W^{w \cdot i} E^i + \begin{pmatrix} -\alpha & q^{w \cdot i\mathsf{T}} \\ q^{w \cdot i} & P^{w \cdot i} \end{pmatrix}$$

$$+ \begin{pmatrix} \beta & 0_{1 \times (d+m)} \\ 0_{(d+m) \times 1} & -\text{Id}_{(d+m) \times (d+m)} \end{pmatrix} \succeq 0$$

(5.50)

Figure 5.12. Summary of generated SDP problem for k-inductive piecewise quadratic invariant for piecewise affine discrete systems.

5.4.3.5 Remark: special case of length 1

When one considers Equations (5.43), (5.46), (5.47), and (5.49) with the set of paths $Paths^1$ of length up to 1, we obtain exactly Equations (5.29), (5.27), and (5.33). In that case, Equation (5.47) does not hold since no nonempty word of length strictly less than 1 exists.

5.4.3.6 Overall method

The algorithm in Figure 5.11 summarizes the method.

5.4.4 Example

The analysis of the running example with increased length generates the following results:

length	$\beta(\sqrt{\beta})$	α	$|Paths^k|$
1	2173 (46.6154)	242.0155	4
2	2133 (46.1844)	233.0847	17
3	1652 (40.6448)	220.8596	73
4	1574 (39.6737)	228.5051	314

Note that the bound α on the piecewise quadratic sublevel applies on different sets of such local Lyapunov functions. Their comparison is meaningless.

5.5 POLYNOMIAL INVARIANTS

5.5.1 Fixpoint expression using polynomial Lyapunov functions

We focus here on a more general family of problems: piecewise polynomial systems. We also rely on more general optimization problems: the cone of positive polynomial and its relaxation/reinforcement as the cone of sum-of-squares polynomials (SOS).

Instead of expressing constraints as linear matrix inequalities (LMI), we can here express constraints as positive polynomial constraints. These constraints will be further reinforced by requiring them to be SOS polynomials.

Let us consider again Equation (5.3) defining inductiveness of computed property with respect to the system semantics.

$$
\begin{cases}
X^{Init} \subseteq P, \\
\forall i \in \mathcal{I}, \ T^i \left(P \cap X^i \right) \subseteq P.
\end{cases}
\tag{5.51}
$$

Encoding property P as the sublevel set of a polynomial p, we obtain the following problem:

$$
\begin{cases}
p(x) \leq 0, & \forall x \in X^{Init}, \\
\forall i \in \mathcal{I}, \ p \left(T^i(x) \right) \leq 0, & \forall x \in P \cap X^i.
\end{cases}
\tag{5.52}
$$

5.5.2 Property-driven analysis

As for the linear case, the previous equation only captures the inductiveness of the sublevel set induced by the polynomial Lyapunov function synthesized. However, no constraint encodes the need to obtain a precise (hence small) invariant. The expressivity of sum-of-square optimization enables us to encode a target property represented as a sublevel set of a polynomial and require the polynomial Lyapunov function to imply this property.

5.5.2.1 Considered properties: sublevel properties $\mathcal{P}_{\kappa,\alpha}$

We restrict our encoding to sublevel properties: those defined as sublevel sets of a given polynomial function.

Definition 5.15 (Sublevel Property).   *Given a polynomial function $\kappa \in \mathbb{R}[x]$ and $\alpha \in \mathbb{R} \cup \{+\infty\}$, we define the sublevel property $\mathcal{P}_{\kappa,\alpha}$ as follows:*

$$\mathcal{P}_{\kappa,\alpha} := \{x \in \mathbb{R}^d \mid \kappa(x) \ll \alpha\}$$

where $\ll$ denotes $\leq$ when $\alpha \in \mathbb{R}$ and denotes $<$ for $+\infty$. The expression $\kappa(x) < +\infty$ expresses the boundedness of $\kappa(x)$ without providing a specific bound α.

Example 5.16 (Sublevel property examples).
Boundedness. *When one wants to bound the reachable values of a system, we can try to bound the l_2-norm of the system: $\mathcal{P}_{\|\cdot\|_2^2, \infty}$ with $\kappa(x) = \|x\|_2^2$. The use of $\alpha = \infty$ does not impose any bound on $\kappa(x)$.*

Safe set. *Similarly, it is possible to check whether a specific bound is matched by either globally using the l_2-norm and a specific α: $\mathcal{P}_{\|\cdot\|_2^2, \alpha}$, or bounding the reachable values of each variable: $\mathcal{P}_{\kappa_i, \alpha_i}$ with $\kappa_i : x \mapsto x_i$ and $\alpha_i \in \mathbb{R}$.*

Avoiding bad regions. *If the bad region can be encoded as a sublevel property $k(x) \leq 0$ then its negation $-k(x) \leq 0$ characterize the avoidance of that bad zone. For example, if one wants to prove that the square norm of the program variables is always greater than 1, then we can consider the property $\mathcal{P}_{\kappa,\alpha}$ with $\kappa(x) = 1 - \|x\|_2^2$ and $\alpha = 0$.*

A sublevel property is called *sublevel invariant* when this property is an inductive invariant of the discrete dynamical system collecting semantics $\mathfrak{C}$. In that case, the sublevel property itself would be an appropriate abstraction of the system. However, this is not the case in general. We rather propose to constrain the search for an inductive polynomial invariant guided by this sublevel property.

5.5.2.2 $\mathcal{P}_{\kappa,\alpha}$-driven inductive polynomial invariant

In this subsection, we explain how we adapt the constraints of Equation (5.52) to compute a d-variate polynomial $p \in \mathbb{R}[x]$ and a bound $w \in \mathbb{R}$, such that the polynomial sublevel sets $P := \{x \in \mathbb{R}^d \mid p(x) \leq 0\}$ and $\mathcal{P}_{\kappa,w}$ satisfy:

$$\mathfrak{C} \subseteq P \subseteq \mathcal{P}_{\kappa,w} \subseteq \mathcal{P}_{\kappa,\alpha} . \tag{5.53}$$

The first (from the left) inclusion forces P to be valid for the whole reachable values set. The second inclusion constrains all elements of P to satisfy the given sublevel property for a certain bound w. The last inclusion requires that the bound w is smaller than the desired level α. When $\alpha = \infty$, any bound w ensures the sublevel property.

We derive sufficient conditions on p and w to satisfy Equation (5.53). Thanks to Equation (5.52), the first inclusion holds: $\mathfrak{C} \subseteq P$.

Now, we are interested in the second and third inclusions at Equation (5.53), that is, the sublevel property satisfaction. The condition $P \subseteq \mathcal{P}_{\kappa,w} \subseteq \mathcal{P}_{\kappa,\alpha}$ can be formulated as follows:

$$\kappa(x) \leq w \leq \alpha, \quad \forall x \in P. \tag{5.54}$$

Recall that we have supposed that P is written as $\{x \in \mathbb{R}^d \mid p(x) \leq 0\}$ where $p \in \mathbb{R}[x]$. Finally, we provide sufficient conditions to satisfy both (5.52) and (5.54). Consider the following optimization problem:

$$\begin{cases} \inf_{p \in \mathbb{R}[x], w \in \mathbb{R}} & w, \ \text{s.t.} \\ & p(x) \leq 0, \quad \forall x \in X^{Init}, \\ \forall i \in \mathcal{I}, p\left(T^i(x)\right) \leq p(x), \quad \forall x \in X^i, \\ & \kappa(x) \leq w + p(x), \quad \forall x \in \mathbb{R}^d. \end{cases} \tag{5.55}$$

Remark that α is not present in Problem (5.55). Indeed, since we minimize w, either there exists a feasible w such that $w \leq \alpha$ and we can exploit this solution or such w is not available and we cannot conclude. However, from Problem (5.55), we can extract (p, w) and in the case where the optimal bound w is greater than α, we could use this solution in conjunction with other abstractions as presented in the following chapter.

Lemma 5.17. *Let (p, w) be any feasible solution of Problem (5.55) with $w \leq \alpha$ or $w < \infty$ in the case of $\alpha = \infty$. Then, (p, w) satisfies both (5.52) and (5.54) with $P := \{x \in \mathbb{R}^d \mid p(x) \leq 0\}$. Finally, P and $\mathcal{P}_{\kappa,w}$ satisfy Equation (5.53).*

In practice, we rely on sum-of-squares programming to solve a relaxed version of Problem (5.55).

5.5.3 SOS relaxed semantics

One way to strengthen the three nonnegativity constraints of Problem (5.55) is to take $\lambda^i = 1$, for all $i \in \mathcal{I}$, $\nu = 1, \alpha = w$, then to consider the *hierarchy* of SOS programs, parametrized by the integer m representing half of the degree of p. All positivity constraints are expressed using sum-of-square polynomials of fixed degrees. As for the k-induction proof, one can increase the degree of such polynomials to consider more general problems. Equation (5.56) details the SOS problem submitted to the solver.

Proposition 5.18. *For a given $m \in \mathbb{N}$, let (p_m, w_m) be any feasible solution of Problem (5.56). Then, (p_m, w_m) is also a feasible solution of Problem (5.55). Moreover, if $w_m \leq \alpha$ then both $P_m := \{x \in \mathbb{R}^d \mid p_m(x) \leq 0\}$ and $\mathcal{P}_{\kappa,w_m}$ satisfy Equation (5.53).*

$$
\begin{cases}
\inf_{p \in \mathbb{R}[x]_{2m}, w \in \mathbb{R}} \quad w, \text{ s.t.} \\[4pt]
-p = \sigma_0 - \sum_{j=1}^{n_{\text{in}}} \sigma_j r_j^{\text{in}}, \\[4pt]
\forall i \in \mathcal{I}, \ p - p \circ T^i = \sigma^i - \sum_{j=1}^{n_i} \mu_j^i r_j^i, \\[4pt]
w + p - \kappa = \psi, \\[4pt]
\forall j = 1, \dots, n_{\text{in}}, \ \sigma_j \in \Sigma[x], \ \deg(\sigma_j r_j^{\text{in}}) \le 2m, \\[4pt]
\sigma_0 \in \Sigma[x], \ \deg(\sigma_0) \le 2m, \\[4pt]
\forall i \in \mathcal{I}, \ \sigma^i \in \Sigma[x], \ \deg(\sigma^i) \le 2m \deg T^i, \\[4pt]
\forall i \in \mathcal{I}, \ \forall j = 1, \dots, n_i, \ \mu_j^i \in \Sigma[x], \\[4pt]
\deg(\mu_j^i r_j^i) \le 2m \deg T^i, \\[4pt]
\psi \in \Sigma[x], \ \deg(\psi) \le 2m,
\end{cases}
\tag{5.56}
$$

where $\forall j \in [1, n_{\text{in}}], r_j^{\text{in}} \le 0$ denotes the initial semialgebraic set, and for all partition i, $\forall j \in [1, n_i], r_j^i \le 0$ denotes the constraints describing the partition.

5.5.4 Examples

Here, we perform some numerical experiments while solving Problem (5.56) (given in Section 4.2.1.3) with several examples. In Section 5.5.4.1, we verify that the program of Example 4.8 satisfies some boundedness property. We also provide examples involving higher dimensional cases. Then, Section 5.5.4.2 focuses on other properties, such as checking that the set of variable values avoids an unsafe region.

We rely on the SDP solver MOSEK to perform the computation.

5.5.4.1 Checking boundedness of the set of variables values

Example 5.19. *Following Example 4.8, we consider the constrained piecewise discrete-time dynamical system* $\mathcal{S} = (X^{Init}, \{X^1, X^2\}, \{T^1, T^2\})$ *with* $X^{Init} = [0.9, 1.1] \times [0, 0.2]$, $X^1 = \{x \in \mathbb{R}^2 \mid r^1(x) \le 0\}$ *with* $r^1 : x \mapsto \|x\|^2 - 1$, $X^2 = \{x \in \mathbb{R}^2 \mid r^2(x) < 0\}$ *with* $r^2 = -r^1$ *and* $T^1 : (x_1, x_2) \mapsto (c_{11}x_1^2 + c_{12}x_2^3, c_{21}x_1^3 + c_{22}x_2^2)$, $T^2 : (x_1, x_2) \mapsto (d_{11}x_1^3 + d_{12}x_2^2, d_{21}x_1^2 + d_{22}x_2^2)$. *We are interested in showing that the boundedness property* $\mathcal{P}_{\|\cdot\|_2^2, \alpha}$ *holds for some positive* α.

Here we illustrate the method by instantiating the program of Example 4.8 with the following input: $a_1 = 0.9$, $a_2 = 1.1$, $b_1 = 0$, $b_2 = 0.2$, $c_{11} = c_{12} = c_{21} = c_{22} = 1$, $d_{11} = 0.5$, $d_{12} = 0.4$, $d_{21} = -0.6$, and $d_{22} = 0.3$. We represent the possible initial values taken by the program variables (x_1, x_2) by picking uniformly N points $(x_1^{(i)}, x_2^{(i)})$ $(i = 1, \dots, N)$ inside the box $X^{Init} = [0.9, 1.1] \times [0, 0.2]$ (see the corresponding square of dots in Figure 5.13). The other dots are obtained after

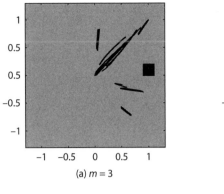

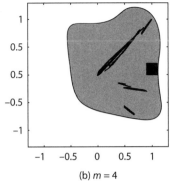

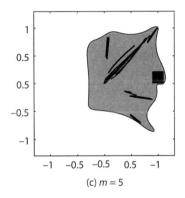

Figure 5.13. A hierarchy of sublevel sets P_m for Example 5.17.

successive updates of each point $(x_1^{(i)}, x_2^{(i)})$ by the program of Example 4.8. The sets of dots in Figure 5.13 are obtained with $N = 100$ and six successive iterations.

At step $m = 3$, Program (5.56) yields a solution $(p_3, w_3) \in \mathbb{R}_6[x] \times \mathbb{R}$ together with SOS certificates, which guarantee the boundedness property, that is, $x \in \mathfrak{C} \implies x \in P_3 := \{p_3(x) \leq 0\} \subseteq \mathcal{P}_{\|\cdot\|_2^2, w_3} \implies \|x\|_2^2 \leq w_3$. One has $p_3(x) := -2.5109$ $02467 - 0.0050x_1 - 0.0148x_2 + 3.0998x_1^2 - 0.8037x_2^3 - 3.0297x_1^3 + 2.5924x_2^2 + 1.52$ $66x_1x_2 - 1.9133x_1^2x_2 - 1.8122x_1x_2^2 + 1.6042x_1^4 + 0.0512x_1^3x_2 - 4.4430x_1^2x_2^2 - 1.89$ $26x_1x_2^3 + 0.5464x_2^4 - 0.2084x_1^5 + 0.5866x_1^4x_2 + 2.2410x_1^3x_2^2 + 1.5714x_1^2x_2^3 - 0.0890$ $x_1x_2^4 - 0.9656x_2^5 + 0.0098x_1^6 - 0.0320x_1^5x_2 - 0.0232x_1^4x_2^2 + 0.2660x_1^3x_2^3 + 0.7746x_1^2$ $x_2^4 + 0.9200x_1x_2^5 + 0.6411x_2^6$ (for the sake of conciseness, we do not display p_4 and p_5).

Figure 5.13 displays in light gray outer approximations of the set of possible values X_1 taken by the program of Example 5.19 as follows: (a) the degree six sublevel set P_3, (b) the degree eight sublevel set P_4, and (c) the degree ten

Table 5.1. Comparison of timing results for Example 5.19.

deg $2m$	# vars	SDP size	time
4	1513	368	$0.82\,s$
6	5740	802	$1.35\,s$
8	15705	1404	$4\,s$
10	35212	2174	$9.86\,s$

Table 5.2. Comparison of timing results for Example 5.20

deg $2m$	# vars	SDP size	time
4	2115	628	$0.84\,s$
6	11950	1860	$2.98\,s$
8	46461	4132	$21.4\,s$
10	141612	7764	$109\,s$

sublevel set P_5. The outer approximation P_3 is coarse as it contains the box $[-1.5, 1.5]^2$. However, solving Problem (5.56) at higher steps yields tighter outer approximations of $\mathfrak{C}$ together with more precise bounds w_4 and w_5 (see the corresponding row in Table 5.4).

We also succeeded in certifying that the same property holds for higher dimensional programs, described in Example 5.20 $(d=3)$ and Example 5.21 $(d=4)$.

Example 5.20. *Here we consider* $X^{Init} = [0.9, 1.1] \times [0, 0.2]^2$, $r^0 : x \mapsto -1$, $r^1 : x \mapsto \|x\|_2^2 - 1$, $r^2 = -r^1$, $T^1 : (x_1, x_2, x_3) \mapsto 1/4(0.8x_1^2 + 1.4x_2 - 0.5x_3^2, 1.3x_1 + 0.5\,x_3^2, 1.4x_2 + 0.8x_3^2)$, $T^2 : (x_1, x_2, x_3) \mapsto 1/4(0.5x_1 + 0.4x_2^2, -0.6x_2^2 + 0.3x_3^2, 0.5x_3 + 0.4x_1^2)$, *and* $\kappa : x \mapsto \|x\|_2^2$.

Example 5.21. *Here we consider* $X^{Init} = [0.9, 1.1] \times [0, 0.2]^3$, $r^0 : x \mapsto -1$, $r^1 : x \mapsto \|x\|_2^2 - 1$, $r^2 = -r^1$, $T^1 : (x_1, x_2, x_3, x_4) \mapsto 0.25(0.8x_1^2 + 1.4x_2 - 0.5x_3^2, 1.3x_1 + 0.5, x_2^2 - 0.8x_4^2, 0.8x_3^2 + 1.4x_4, 1.3x_3 + 0.5x_4^2)$, $T^2 : (x_1, x_2, x_3, x_4) \mapsto 0.25(0.5x_1 + 0.4x_2^2, -0.6x_1^2 + 0.3x_2^2, 0.5x_3 + 0.4x_4^2, -0.6x_3 + 0.3x_4^2)$, *and* $\kappa : x \mapsto \|x\|_2^2$.

Tables 5.1, 5.2, and 5.3 report several data obtained while solving Problem (5.56) at step m, $(2 \leq m \leq 5)$, either for Example 5.19, Example 5.20, or Example 5.21. Each instance of Problem (5.56) is recast as an SDP program, involving a total number of "# vars" SDP variables, with an SDP matrix of size "SDP size." We indicate the CPU time required to compute the optimal solution of each SDP program with MOSEK.

Table 5.3. Comparison of timing results for Example 5.21

deg $2m$	# vars	SDP size	time
4	7202	1670	$2.85\ s$
6	65306	6622	$57.3\ s$
8	18480	373057	$1534\ s$
10	–	–	–

The symbol "−" means that the corresponding SOS program could not be solved within one day of computation. These benchmarks illustrate the computational considerations mentioned in Section 4.2.1.3 as it takes more CPU time to analyze higher dimensional programs. Note that it is not possible to solve Problem (5.56) at step 5 for Example 5.21. A possible workaround to limit this computational blow-up would be to exploit the sparsity of the system.

5.5.4.2   Other properties

Here we consider the program given in Example 5.22. One is interested in showing that the set X_1 of possible values taken by the variables of this program does not meet the ball B of center $(-0.5, -0.5)$ and radius 0.5.

Example 5.22.   *Let consider the piecewise polynomial system* $\mathcal{S} = (X^{Init}, \{X^1, X^2\}, \{T^1, T^2\})$ *with* $X^{Init} = [0.5, 0.7] \times [0.5, 0.7]$, $X^1 = \{x \in \mathbb{R}^2 \mid r^1(x) \leq 0\}$ *with* $r^1 : x \mapsto \|x\|_2^2 - 1$, $X^2 = \{x \in \mathbb{R}^2 \mid r^2(x) \leq 0\}$ *with* $r^2 = -r^1$ *and* $T^1 : (x_1, x_2) \mapsto (x_1^2 + x_2^3, x_1^3 + x_2^2)$, $T^2 : (x, y) \mapsto (0.5x_1^3 + 0.4x_2^2, -0.6x_1^2 + 0.3x_2^2)$. *With* $\kappa : (x_1, x_2) \mapsto 0.25 - (x_1 + 0.5)^2 - (x_2 + 0.5)^2$, *one has* $B := \{x \in \mathbb{R}^2 \mid 0 \leq \kappa(x)\}$. *Here, one shall prove* $x \in \mathfrak{C} \implies \kappa(x) < 0$ *while computing some negative* α *such that* $\mathfrak{C} \subseteq \mathcal{P}_{\kappa, \alpha}$. *Note that* κ *is not a norm, by contrast with the previous examples.*

At step $m = 3$ (resp. $m = 4$), Program (5.56) yields a nonnegative solution w_3 (resp. w_4). Hence, it does not allow us to certify that $\mathfrak{C} \cap B$ is empty. This is illustrated in both Figure 5.14 (a) and Figure 5.14 (b), where the light gray region does not avoid the ball B. However, solving Program (5.56) at step $m = 5$ yields a negative bound w_5 together with a certificate that $\mathfrak{C}$ avoids the ball B (see Figure 5.14 (c)). The corresponding values of w_m ($m = 3, 4, 5$) are given in Table 5.4.

Finally, one analyzes the program given in Example 5.23.

Example 5.23.   *(adapted from Example 3 in [119])*
Let $\mathcal{S}$ *be the piecewise polynomial system* $(X^{Init}, \{X^1, X^2\}, \{T^1, T^2\})$ *with* $X^{Init} = [-1, 1] \times [-1, 1]$, $X^1 = \{x \in \mathbb{R}^2 \mid r^1(x) \leq 0\}$ *with* $r^1 : x \mapsto x_2 - x_1$, $X^2 = \{x \in \mathbb{R}^2 \mid r^2(x) \leq 0\}$ *with* $r^2 = -r^1$, *and* $T^1 : (x_1, x_2) \mapsto (0.687x_1 + 0.558x_2 - 0.0001 * x_1x_2, -0.292x_1 + 0.773x_2)$, $T^2 : (x, y) \mapsto (0.369x_1 + 0.532x_2 - 0.0001x_1^2$,

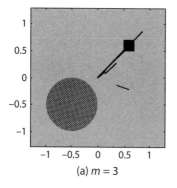

(a) $m = 3$

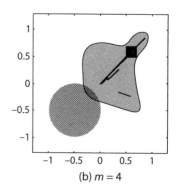

(b) $m = 4$

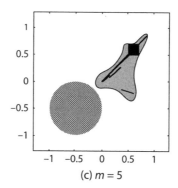

(c) $m = 5$

Figure 5.14. A hierarchy of sublevel sets P_m for Example 5.20.

Table 5.4. Hierarchies of bounds obtained for various properties

	Ex. 5.19	Ex. 5.22	Ex. 5.23	
κ	$\|\cdot\|_2^2$	κ	$\|\cdot\|_2^2$	κ_2
w_2	639	0.25	10.2	5.66
w_3	17.4	0.249	2.84	2.81
w_4	2.44	0.0993	2.84	2.78
w_5	2.02	-0.0777	2.84	2.78

Note: Ex.5.22 $\kappa = x \mapsto 0.25 - \|x + 0.5\|_2^2$ and Ex. 5.23 $\kappa_2 = x \mapsto \|T^1(x) - T^2(x)\|_2^2$.

$-1.27x_1 + 0.12x_2 - 0.0001x_1x_2)$. *We consider the boundedness property* $\kappa_1 :=$ $\|\cdot\|_2^2$ *as well as* $\kappa_2(x) := \|T^1(x) - T^2(x)\|_2^2$. *The function* κ_2 *can be viewed as the absolute error made by updating the variable* x *after a possibly "wrong" branching. Such behaviors could occur while computing wrong values for the conditionals (e.g.,* r^1) *using floating-point arithmetics. Table 5.4 indicates the hierarchy of bounds obtained after solving Problem (5.56) with* $m = 3, 4, 5$, *for both properties. The bound* $w_5 = 2.84$ *(for* κ_1) *implies that the set of reachable values may not be included in the initial set* X^{Init}. *A valid upper bound of the error function* κ_2 *is given by* $w_5 = 2.78$.

5.6 IMAGE MEASURE METHOD

Let us consider now another approach inspired by the works by Henrion, Lasserre, Magron, and Korda [115, 123, 124, 134, 142], bounding precisely the reachable states, $\mathfrak{C}$ of a dynamical system. Assuming this set $\mathfrak{C}$ lives in a given compact set $\mathbf{X}$, let us denote, in the following, by $\mathbf{X}^{(\infty)}$ the set of reachable states restricted to $\mathbf{X}$:

$$\mathbf{X}^{(\infty)} := \{(\mathbf{x}_t)_{t \in \mathbb{N}} \subseteq \mathbf{X} : \mathbf{x}_{t+1} = f(\mathbf{x}_t), \forall t \in \mathbb{N}, \mathbf{x}_0 \in \mathbf{X}_0\}.$$

The idea is to express $\mathbf{X}^{(\infty)}$ as a minimization optimization problem in which we search for an inductive sublevel set, containing the initial set, and in which the volume, with respect to Lebesgue measure, is minimal. Thanks to the compactness of $\mathbf{X}$, the Lebesgue measure is defined. Furthermore, when choosing an appropriate compact set X, for example, a hypercube or a ball, the computation of the volume of a semialgebraic set is expressible in a linear fashion, over the moments associated to monomials.

This method is inspired by a long line of works manipulating polynomial systems properties and compact sets. In [87], the authors addressed the problem of computing over-approximations of the volume of a general basic compact semialgebraic set, described by the intersection of a finite number of polynomial superlevel sets, whose coefficients are known in advance. Further work focused on over-approximating semialgebraic sets where such a description is not explicitly known: in [141], the author derives converging outer (resp. inner) approximations of sets defined with existential (resp. universal) quantifiers; in [142], the authors approximate the image set of a compact semialgebraic set S under a polynomial map f. The current study can be seen as an extension of [142], when S stands for the set of initial conditions, f represents the dynamics of a discrete-time system, and only one iteration is performed from S.

One can characterize a hierarchy of converging convex approximations derived from an infinite-dimensional linear programming (LP) reformulation of the problem. Through moment relaxations of this LP, and characterization on the dual problem, one can compute tight over-approximations of the reachable set.

5.6.1   Primal: maximizing measure support

The initial expression of the problem relies on the characterization of the indicator function $\mathbf{1}_{\mathfrak{C}}$ of the set $\mathfrak{C}$.

$$\mathbf{1}_{\mathbf{A}}(\mathbf{x}) := \begin{cases} 1 & \text{if } \mathbf{x} \in \mathbf{A}, \\ 0 & \text{otherwise}. \end{cases}$$

Let us introduce some definitions.

Definition 5.24 (Borel Measures Vector Space). *Given a compact set $\mathbf{A} \subset \mathbb{R}^n$, let us denote by $\mathcal{M}(\mathbf{A})$ the vector space of finite signed Borel measures supported on $\mathbf{A}$, namely real-valued functions from the Borel sigma algebra $\mathcal{B}(\mathbf{A})$.*

Definition 5.25 (Measure Support). *The support of a measure $\mu \in \mathcal{M}(\mathbf{A})$ is defined as the set of all points $\mathbf{x}$ such that for each open neighborhood $\mathbb{B}$ of $\mathbf{x}$, one has $\mu(\mathbb{B}) > 0$. Note that this set is closed by construction.*

Definition 5.26 (Lebesgue Measure on a Subset). *The restriction of the Lebesgue measure on a subset $\mathbf{A} \subseteq \mathbf{X}$ is $\lambda_{\mathbf{A}}(d\mathbf{x}) := \mathbf{1}_{\mathbf{A}}(\mathbf{x}) \, d\mathbf{x}.$, where $\mathbf{1}_{\mathbf{A}} : \mathbf{X} \to \{0,1\}$ stands for the indicator function on $\mathbf{A}$.*

Definition 5.27 (Moments of Lebesgue Measure). *The moments of the Lebesgue measure on $\mathbf{X}$ are denoted by*

$$y_{\beta}^{\mathbf{X}} := \int_{\mathbf{X}} \mathbf{x}^{\beta} \lambda_{\mathbf{X}}(d\mathbf{x}) \in \mathbb{R}, \quad \beta \in \mathbb{N}^n. \tag{5.57}$$

Definition 5.28 (Lebesgue Volume). *The Lebesgue volume of $\mathbf{X}$ is defined by $\operatorname{vol} \mathbf{X} := y_0^{\mathbf{X}} = \int_{\mathbf{X}} \lambda_{\mathbf{X}}(d\mathbf{x})$.*

Definition 5.29 (Image Measure). *Given a positive measure $\mu \in \mathcal{M}_+(\mathbf{X})$, the so-called pushforward measure (or image measure, see, e.g., [31, Section 1.5]) of μ under f is defined as follows:*

$$f_{\#}\mu(\mathbf{A}) := \mu(f^{-1}(\mathbf{A})) = \mu(\{\mathbf{x} \in \mathbf{X} : f(\mathbf{x}) \in \mathbf{A}\}),$$

for every set $\mathbf{A} \in \mathcal{B}(\mathbf{X})$.

Definition 5.30 (Invariant Measures). *A measure μ is invariant w.r.t. f when μ satisfies $\mu = f_{\#}\mu$.*

The set $\mathbf{X}^{\mathrm{inv}}$ is defined as the union of supports of all invariant measures w.r.t. f being dominated w.r.t. $\lambda_{\mathbf{X}}$ (the restriction of Lebesgue measure over $\mathbf{X}$).

Lemma 5.31. *For any given $T \in \mathbb{N}_0$, $\alpha > 1$ and a measure $\mu_0 \in \mathcal{M}(\mathbf{X}_0)$, there exist measures $\mu_T, \nu \in \mathcal{M}(\mathbf{X})$ which satisfy the discrete Liouville's Equation:*

$$\mu_T + \nu = \alpha f_{\#}\nu + \mu_0. \tag{5.58}$$

Using Liouville equation that encodes a sort of certain conservation law for measure supports, one can derive the following primal formulation. To approximate the set $\mathbf{X}^* := \mathbf{X}^{\text{inv}} \cup \mathbf{X}^{(\infty)}$, one considers the infinite-dimensional linear programming (LP) problem, for a given $\alpha > 1$:

$$
\begin{aligned}
p^* := \sup_{\mu_0, \mu, \hat{\mu}, \nu} \quad & \int_{\mathbf{X}} \mu \\
\text{s.t.} \quad & \mu + \hat{\mu} = \lambda_{\mathbf{X}}, \\
& \mu + \nu = \alpha f_{\#}\nu + \mu_0, \\
& \mu_0 \in \mathcal{M}_+(\mathbf{X}_0), \quad \mu, \hat{\mu}, \nu \in \mathcal{M}_+(\mathbf{X}).
\end{aligned}
\tag{5.59}
$$

Intuitively, μ denotes the measure of terminal reachable states. However, since the trace is not bounded by the equation, it can denote any reachable state. The second constraint is the so-called Liouville equation: it encodes the system semantics within the constraints.

Remark 5.32 (Limitation). *Note that since $X^* \supset X_\infty$ the computation of X^* also contains X^{inv} the set describing measure invariant sets. This will add as reachable points unfeasible ones. This can range from nonreachable fixpoints, co-limit cycles, or strange attractors. As an example, if one considers the simple linear system that rotates its input without contraction, the reachable state space is the closure by rotation of the initial one. But, with the presented method, if $X^0 = \|x\|_2^2 < .5$ while $X = \|x\|_2^2 < 1$ then, by rotation, we have $X^\infty = X^0$ while the method selects all X.*

5.6.2 Dual: minimizing positive functions

Positive measures are not fitted with a scalar product and are then not Hilbert spaces. The (pre-)dual of positive measures is the set of positive continuous functions $\mathcal{C}(\mathbf{X})$. Using the elements of duality introduced in Sec. 4.2.3, one can construct the dual problem of the maximization of measure support:

$$
\begin{aligned}
d^* := \inf_{v, w} \quad & \int w(\mathbf{x}) \, \lambda_{\mathbf{X}}(d\mathbf{x}) \\
\text{s.t.} \quad & v(\mathbf{x}) \geq 0, \quad \forall \mathbf{x} \in \mathbf{X}_0, \\
& w(\mathbf{x}) \geq 1 + v(\mathbf{x}), \quad \forall \mathbf{x} \in \mathbf{X}, \\
& w(\mathbf{x}) \geq 0, \quad \forall \mathbf{x} \in \mathbf{X}, \\
& \alpha \, v(f(\mathbf{x})) \geq v(\mathbf{x}), \quad \forall \mathbf{x} \in \mathbf{X}, \\
& v, w \in \mathcal{C}(\mathbf{X}).
\end{aligned}
\tag{5.60}
$$

Intuitively, we are interested in the superlevel set of the function $v(x)$. $v(x) \geq 0$ on initial state, and this positive is preserved along system trajectories: this is encoded by the constraint $\alpha \, v(f(\mathbf{x})) \geq v(\mathbf{x})$. If a state x is reachable, then

$v(x) \geq 0$ and s does its successor $v(f(\mathbf{x})) \geq v(\mathbf{x}) \geq 0$. However, v can be anything outside reachable states. The positive function w is such that when $v(x) > 0$ then $w(x)$ is above a specific threshold, here 1. And since $w(x)$ is positive over X, one can minimize its volume.

5.6.3 Hierarchy of abstractions

Using the Henrion and Lasserre approach [39, 115, 123, 134], we abstract positive functions by SOS polynomials. Thanks to theoretical results, the method converges in volume towards X^*.

Positivity of polynomial expressions under certain semialgebraic constraints is ensured by imposing them to be an SOS polynomial of a given degree $2m$, as we saw in Section 5.5.3, for example, in Eq. (5.56). The problem to solve becomes:

$$
d_r^* := \inf_{v,w} \quad \sum_{\beta \in \mathbb{N}_{2r}^n} w_\beta z_\beta^{\mathbf{X}}
$$

$$
\text{s.t.} \quad v = \sigma_0 - \sum_{j=1}^{n_{in}} \sigma_j^0 r_j^{in} ,
$$

$$
w - 1 - v = \sigma_1 - \sum_{j=1}^{n_X} \sigma_j^1 r_j^X ,
$$

$$
\alpha \, v \circ f - v = \sigma_2 - \sum_{j=1}^{n_X} \sigma_j^2 r_j^X ,
$$

(5.61)

$$
w = \sigma_3 - \sum_{j=1}^{n_X} \sigma_j^3 r_j^X ,
$$

$$
v, w \in \mathbb{R}_{2m}[\mathbf{x}].
$$

$$
\forall i \in \{0, 1, 2, 3\}, \sigma_i \in \Sigma[x] \ , \ \deg(\sigma_i) \leq 2m \ ,
$$

$$
\forall j = 1, \dots, n_{in} \ , \ \sigma_j \in \Sigma[x] \ , \ \deg(\sigma_j r_j^{in}) \leq 2m \ ,
$$

$$
\forall j = 1, \dots, n_X, i \in \{1, 2, 3\} \ , \ \sigma_j^i \in \Sigma[x] \ ,
$$

$$
\deg(\sigma_j^i r_j^{in}) \leq 2m \ ,
$$

where, as in 5.5.3, initial state belong to the semialgebraic set

$$
X^0 = \left\{ x \ \middle| \ \bigwedge_{j \in [1, n_{in}]} r_j^{in}(x) \leq 0 \right\} ,
$$

and the compact set X is defined as $\left\{ x \ \middle| \ \bigwedge_{j \in [1, n_X]} r_j^X(x) \leq 0 \right\}$.

One of the key aspects is the capability to express the volume of the zero superlevel set $w(x) \geq 0$ within the compact $\mathbf{X}$: $\int w(\mathbf{x}) \lambda_{\mathbf{X}}(d\mathbf{x})$. Thanks to [39], when w is of fixed degree and $\mathbf{X}$ has a simple shape such as a unit ball, one can pre-compute the moment $z_\beta^{\mathbf{X}} = \int_{\mathbf{X}} \beta(x) dx$ associated to each monomial β and compute the simpler $\sum_{\beta \in \mathbb{N}_{2r}^n} w_\beta z_\beta^{\mathbf{X}}$ where w_β denotes the coefficient associated to the monomial β in w.

5.6.4 Experiments

5.6.4.1 Toy example

First, let us consider the made-up discrete-time polynomial system defined by

$$x_1^+ := \frac{1}{2}(x_1 + 2x_1 x_2),$$
$$x_2^+ := \frac{1}{2}(x_2 - 2x_1^3),$$

with initial state constraints $\mathbf{X}_0 := \{\mathbf{x} \in \mathbb{R}^2 : (x_1 - \frac{1}{2})^2 + (x_2 - \frac{1}{2})^2 \leq \frac{1}{4^2}\}$ and general state constraints within the unit ball $\mathbf{X} := \{\mathbf{x} \in \mathbb{R}^2 : \|x\|_2^2 \leq 1\}$. In Figure 5.15, we represent in plain gray the outer approximations $\mathbf{X}^r$ of $\mathbf{X}^*$ obtained by the method, for increasing values of the relaxation order r (from $r = 4$ to $r = 14$). In each figure, the set of points are obtained by simulation for the first 7 iterates. More precisely, each could of points corresponds to (under approximations of) the successive image sets $\mathbf{X}_1, \ldots, \mathbf{X}_7$ of the points obtained by uniform sampling of $\mathbf{X}_0$ under $f, \ldots, f^7$ respectively. The set $\mathbf{X}_0$ is lighter, in the upper right part of the pictures, and the set $\mathbf{X}_7$ is darker, near the center. The dotted circle represents the boundary of the unit ball $\mathbf{X}$. Figure 5.15 shows that the over-approximations are already quite tight for low degrees.

5.6.4.2 FitzHugh-Nagumo neuron model

Here we consider the discretized version (taken from [106, Section 5]) of the FitzHugh-Nagumo model [2], which was originally a continuous-time polynomial system modeling the electrical activity of a neuron:

$$x_1^+ := x_1 + 0.2(x_1 - x_1^3/3 - x_2 + 0.875),$$
$$x_2^+ := x_2 + 0.2(0.08(x_1 + 0.7 - 0.8x_2)),$$

with initial state constraints $\mathbf{X}_0 := [1, 1.25] \times [2.25, 2.5]$ and general state constraints $\mathbf{X} := \{\mathbf{x} \in \mathbb{R}^2 : (\frac{x_1 - 0.1}{3.6})^2 + (\frac{x_2 - 1.25}{1.75})^2 \leq 1\}$. Figure 5.16 illustrates that the over-approximations provide useful indications on the system behavior, in particular for higher values of r. Indeed, $\mathbf{X}^{10}$ and $\mathbf{X}^{12}$ capture the presence of the central "hole" made by periodic trajectories and $\mathbf{X}^{14}$ shows that there is

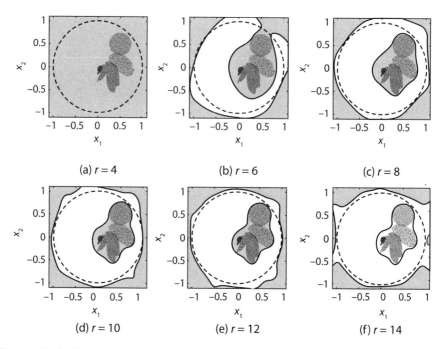

Figure 5.15. Outer approximations $\mathbf{X}^r$ (plain gray) of $\mathbf{X}^*$ (dot samples) for Example 5.6.4.1, from $r = 4$ to $r = 14$.

a gap between the first discrete-time steps and the iterations corresponding to these periodic trajectories.

5.7 RELATED WORKS

Automated nonlinear analyses are not very common in formal verification. A line of works [44, 50] rely on iterative computations using Gröbner bases to synthesize polynomial equality invariants. Similar to Karr's domain representing affine relationships among variables [7], these domains extract polynomial relationships between variables. More recent work [131] relies on a kind of weakest precondition computation to synthesize these polynomial equalities. All these works cannot, in the current state, express semialgebraic sets nor capture the stability of a linear controller.

Regarding quadratic invariants, i.e., ellipsoids, they are not fitted with a lattice structure since there is no unique smallest ellipsoids containing a set of ellipsoids. Unrolling techniques such as [47, 56, 73, 130] enable the precise analysis of linear systems by solving mathematically these dynamical systems. While more precise than the approach we proposed, these techniques can hardly

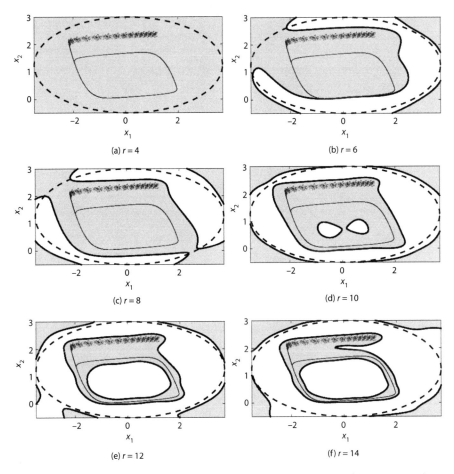

Figure 5.16. Outer approximations $\mathbf{X}^r$ (plain gray) of $\mathbf{X}^*$ (dot samples) for Example 5.6.4.2, from $r = 4$ to $r = 14$.

handle disjunction or saturations. Their use could however be used locally to improve the precision of the analyses. The use of convex optimization such as SDP or SOS to synthesize sublevel set properties was proposed by Cousot in [53] providing simple inductive templates and without addressing methods to check the soundness of the result. Other recent approaches such as [143] proposed a classical Kleene iterations-based abstract domain for ellipsoids in which the join operator is implemented as the call to an SDP solvers synthesizing the minimal volume ellipsoid. The interesting approach of [139] proposed an algebraic method to manipulate a specific class of ellipsoids (zero-centered ellipsoids), without the need to use a numerical tool such as an SDP solver to compute such minimal volume ellipsoids.

A last related category of analyses is the computation of nonconvex properties. Nonconvex properties were also used to express disjunctions as holes in a given more classical convex abstraction: difference-bound matrices (DBM), the underlying domain of octagons, with disequality constraints [74], or the Donut domain [111].

Chapter Six

Template-based Analyses and Min-policy Iteration

WHILE THE previous chapter addressed the direct synthesis of invariants as bound templates, there are other configurations in which we are interested in bounding provided templates.

A first case arises when the previous method, as in Equations (5.10) and (5.11), only synthesizes the template but not the bound. A second appears when one wants to analyze a system with multiple templates. Typically, we are interested in bounds on each variable and want to consider the templates $p(x) = x_i^2$ for each variable x_i in state characterization $x \in \Sigma$. The current chapter proposes a policy iteration algorithm, based on SOS optimization, to refine such template bounds. In practice, we use it by combining a Lyapunov based template obtained using one of the previous methods with additional templates encoding bounds on some variables or property specific templates.

6.1 TEMPLATE-BASED ABSTRACT DOMAINS

Let us now assume that the abstraction is based on a template abstraction. We recall that a template is a real-valued function $p : \Sigma \to \mathbb{R}$. For the rest of the chapter, we assume that these templates are given.

For each template p, one can characterize an abstract domain $D_p^\#$ as presented in Sec. 2.5. We also denote by $\bar{\mathbb{R}} = \mathbb{R} \cup \{-\infty, +\infty\}$ the extension of $\mathbb{R}$ with infinite values and by $\dot{\leq}$ the extension of $\leq$ to those values.

As for the characterization of the fixpoint presented earlier, this abstraction also defines a complete lattice. The order relation $\dot{\leq}$ is total and relies on the real number order applied to the level sets. The join and meet of two abstract values, i.e., the two scalars representing sublevel sets, are computed with max and min.

$$D_p^\# = \langle \bar{\mathbb{R}}, \dot{\leq}, \max, \min, -\infty, +\infty \rangle$$

The abstraction and concretization functions are defined as:

$$\alpha_p : S \mapsto \max\{p(s) | s \in S\}$$
$$\gamma_p : \lambda \mapsto \{s \in S | p(s) \dot{\leq} \lambda\}.$$

Multiple templates could be considered at once. Let $\mathbb{P}$ be a finite family of templates $(p_i)_{0 \leq i < n}$ and $\mathcal{F}\left(\mathbb{P}, \overline{\mathbb{R}}\right)$ be the set of functions from $\mathbb{P}$ to $\overline{\mathbb{R}} = \mathbb{R} \cup \{-\infty, +\infty\}$. We fit $\mathcal{F}\left(\mathbb{P}, \overline{\mathbb{R}}\right)$ with the functional partial order $\leq_{\mathbb{F}}$, i.e., $v \leq_{\mathbb{F}} w$ iff $v(p) \leq w(p)$ for all $p \in \mathbb{P}$. This defines our abstract domain, the lattice

$$D_{\mathbb{P}}^{\#} = \langle \mathcal{F}\left(\mathbb{P}, \overline{\mathbb{R}}\right), \leq_{\mathbb{F}}, \max_{\mathbb{F}}, \min_{\mathbb{F}}, (-\infty)_{\mathbb{F}}, (+\infty)_{\mathbb{F}} \rangle$$

where the functions $\max_{\mathbb{F}}, \min_{\mathbb{F}}$ are lift of max and min to functions. $(\pm\infty)_{\mathbb{F}}$ denote the functions $p \in \mathbb{P} \mapsto \pm\infty$.

We characterize the abstraction $\star$ and concretization $\dagger$ functions. Let $w \in \mathcal{F}\left(\mathbb{P}, \overline{\mathbb{R}}\right)$ and $X \in \mathbb{R}^n$. The concretization of w to sets gives the set $w^{\star}$:

$$w^{\star} = \{x \in \mathbb{R}^d \mid p(x) \leq w(p), \forall\, p \in \mathbb{P}\}. \tag{6.1}$$

While the abstraction of X to $\mathcal{F}\left(\mathbb{P}, \overline{\mathbb{R}}\right)$ is defined by the abstract element $X^{\dagger}$:

$$X^{\dagger}(p) := \sup_{x \in X} p(x). \tag{6.2}$$

6.2 TEMPLATE ABSTRACTION FIXPOINT AS AN OPTIMIZATION PROBLEM

Let us summarize the current definitions:

- The collecting semantics of a system is defined using Equation (2.17) as the least fixpoint of an isomorphism over a set of states and is characterized by the minimum set $S \in \wp(\Sigma)$ of the postfixpoints $F(S) \subseteq S$.
- A possible set of abstraction is defined by templates-based abstract domains. An abstract domain is specified by a finite family of templates, i.e., a real-valued function over system states. An abstract value is a vector of real values characterizing sublevel sets of templates.

Then computing an inductive invariant in the templates domain boils down to providing, for each template p, a bound $w(p)$ such that the intersection over the templates p of sublevel sets $\{x \in \mathbb{R}^d \mid p(x) \leq w(p)\}$ is an inductive invariant. We recall that, in our context, a template is simply an a priori fixed multivariate polynomial.

We need to express the inductiveness of the sets $w^{\star}$ into inequalities on w. By definition the set $w^{\star}$ is an inductive invariant iff $F(w^{\star}) \subseteq w^{\star}$, that is:

$$\bigcup_{i \in \mathcal{I}} T^i(w^{\star} \cap X^i) \cup X^{Init} \subseteq w^{\star}.$$

By definition, $w^\star$ is an inductive invariant iff:

$$\forall p \in \mathbb{P}, \ \forall x \in \bigcup_{i \in \mathcal{I}} T^i(w^\star \cap X^i) \cup X^{Init}, \ p(x) \leq w(p).$$

Using the definition of the supremum, $w^\star$ is an inductive invariant iff:

$$\forall p \in \mathbb{P}, \quad \sup_{x \in \bigcup_{i \in \mathcal{I}} T^i(w^\star \cap X^i) \cup X^{Init}} p(x) \leq w(p).$$

Now, consider $p \in \mathbb{P}$. Using the fact that for all $A, B \subseteq \mathbb{R}^d$ and for all functions f, $\sup_{A \cup B} f = \sup\{\sup_A f, \sup_B f\}$:

$$\sup_{x \in \bigcup_{i \in \mathcal{I}} T^i(w^\star \cap X^i) \cup X^{Init}} p(x) =$$

$$\sup\left\{\sup_{i \in \mathcal{I}} \sup_{x \in T^i(w^\star \cap X^i)} p(x), \ \sup_{x \in X^{Init}} p(x)\right\}.$$

By definition of the image:

$$\sup_{x \in \bigcup_{i \in \mathcal{I}} T^i(w^\star \cap X^i) \cup X^{Init}} p(x) =$$

$$\sup\left\{\sup_{i \in \mathcal{I}} \sup_{y \in w^\star \cap X^i} p(T^i(y)), \ \sup_{x \in X^{Init}} p(x)\right\}.$$

Let us introduce the following notation to denote the image of a set $w^\star$ by a guarded update function T^i, for all $p \in \mathbb{P}$:

$$F_i^\sharp(w)(p) := \sup_{x \in w^\star \cap X^i} p(T^i(x)).$$

We also recall the definition of abstraction applied on initial state:

$$X^{Init^\dagger}(p) := \sup_{x \in X^{Init}} p(x).$$

Finally, we define the function from $\mathcal{F}\left(\mathbb{P}, \overline{\mathbb{R}}\right)$ to itself, for all $w \in \mathcal{F}\left(\mathbb{P}, \overline{\mathbb{R}}\right)$:

$$F^\sharp(w) := \sup\left\{\sup_{i \in \mathcal{I}} F_i^\sharp(w), X^{Init^\dagger}\right\}.$$

By construction, we obtain the following proposition:

Proposition 6.1. *Let $w \in \mathcal{F}\left(\mathbb{P}, \overline{\mathbb{R}}\right)$. Then $w^{\star}$ is an inductive invariant (i.e., $F(w^{\star}) \subseteq w^{\star}$) iff $F^{\sharp}(w) \leq_{\mathbb{F}} w$.*

From Prop. 6.1, $\inf\{w \in \mathcal{F}\left(\mathbb{P}, \overline{\mathbb{R}}\right)^{n} \mid F^{\sharp}(w) \leq_{\mathbb{F}} w\}$ identifies the smallest inductive invariant $w^{\star}$ of the form (6.1).

Example 6.2. *Let us consider the system defined in Example 5.23. Let us consider the same templates basis $\mathbb{P} = \{q_1, q_2, p\}$ where $q_1(x) = x_1^2$, $q_2(x) = x_2^2$, and p is a well-chosen polynomial of degree 6. Let $w \in \mathcal{F}\left(\mathbb{P}, \overline{\mathbb{R}}\right)$. For $i = 1$ and the templates q_1, we have:*

$$F_1^{\sharp}(w)(q_1) =$$
$$\sup_{\substack{-x_1^2 + 1 \leq 0 \\ x_1^2 \leq w(q_1), \\ x_2^2 \leq w(q_2), \\ p(x) \leq w(p).}} (0.687x_1 + 0.558x_2 - 0.0001x_1x_2)^2$$

Indeed, $X^1 = \{x \in \mathbb{R}^2 \mid -x_1^2 + 1\}$ and the dynamics associated with X^1 is the polynomial function T^1 defined for all $x \in \mathbb{R}^2$ by: $T^1(x) = \binom{0.687x_1 + 0.558x_2 - 0.0001x_1x_2}{-0.292x_1 + 0.773x_2}$ and thus since q_1 computes the square of the first coordinates $q_1(T^1(x)) = (0.687x_1 + 0.558x_2 - 0.0001x_1x_2)^2$.

With $w \in \mathcal{F}\left(\mathbb{P}, \overline{\mathbb{R}}\right)$, computing $F^{\sharp}(w)$ boils down to solving a finite number of nonconvex polynomial optimization problems. General methods do not exist to solve such problems. In Section 6.3, we propose a method based on sum-of-squares (SOS) to over-approximate $F^{\sharp}(w)$.

6.3   SOS-RELAXED SEMANTICS

In this section, we introduce the relaxed functional on which we will compute a fixpoint, yielding a further over-approximation of the set $\mathcal{R}$ of reachable values. This relaxed functional is constructed from a Lagrange relaxation of maximization problems involved in the evaluation of $F^{\sharp}$ and sum-of-squares strengthening of polynomial nonnegativity constraints.

6.3.1   Relaxed semantics

The computation of $F^{\sharp}$ as a polynomial maximization problem cannot be directly performed using numerical solvers. We use the SOS reinforcement mechanisms described above to relax the computation and characterize an abstraction of $F^{\sharp}$.

We still assume the knowledge of the template basis $\mathbb{P}$, involving polynomials of degree at most $2m$. Let us define $\mathcal{F}\left(\mathbb{P}, \mathbb{R}_{+}\right)$ the set of nonnegative functions over $\mathbb{P}$, i.e., $g \in \mathcal{F}\left(\mathbb{P}, \mathbb{R}_{+}\right)$ iff for all $p \in \mathbb{P}$, $g(p) \in \mathbb{R}_{+}$. Let $p \in \mathbb{P}$ and $w \in \mathcal{F}\left(\mathbb{P}, \overline{\mathbb{R}}\right)$. Starting from the definition of $F_i^{\sharp}$, one obtains the following:

$$\left(F_i^\sharp(w)\right)(p)$$

$$= \sup_{\substack{q(x)\leq w(q),\ \forall q\in\mathbb{P}\\ r_j^i(x)\leq 0,\ \forall j\in\mathbf{In}_i}} p(T^i(x))$$

$$\leq \inf_{\substack{\lambda\in\mathcal{F}(\mathbb{P},\mathbb{R}_+)\\ \sigma\in\Sigma[x],\mu_l\in\Sigma[x]\\ \deg(\sigma)\leq 2m\deg T^i\\ \deg(\mu_l r_l^i)\leq 2m\deg T^i}} \sup_{x\in\mathbb{R}^d} p(T^i(x))\ +\ \sum_{q\in\mathbb{P}}\lambda(q)(w(q)-q(x))$$

$$-\sum_{l=1}^{n^i}\mu_l(x)r_l^i(x)$$

$$\leq \inf_{\lambda,\sigma,\mu_l,\eta} \eta$$

$$\text{s.t.}\left\{\begin{array}{l} \eta - p\circ T^i - \displaystyle\sum_{q\in\mathbb{P}}\lambda(q)(w(q)-q)\\[2ex] \qquad\qquad + \displaystyle\sum_{l=1}^{n^i}\mu_l r_l^i = \sigma,\\[2ex] \lambda\in\mathcal{F}\left(\mathbb{P},\mathbb{R}_+\right),\ \sigma\in\Sigma[x],\\ \mu_l\in\Sigma[x],\ \eta\in\mathbb{R},\\ \deg(\sigma)\leq 2m\deg T^i,\\ \deg(\mu_l r_l^i)\leq 2m\deg T^i \end{array}\right.$$

(using an SOS reinforcement to remove the sup).

We denote by $\Sigma[x]^n$ the set of n-tuples of SOS polynomials. For clarity purposes, the dependency on i is omitted within the notations of the multipliers μ_l. Moreover, let us write $\sum_{l=1}^{n^i}\mu_l r_l^i$ as $\langle\mu,r^i\rangle$. Finally, we write $\left(F_i^{\mathcal{R}}(w)\right)(p)$ the over-approximation of $\left(F_i^\sharp(w)\right)(p)$, defined as follows:

$$\left(F_i^{\mathcal{R}}(w)\right)(p) = \inf_{\lambda,\sigma,\mu,\eta}\ \eta$$

$$\text{s.t.}\left\{\begin{array}{l} \eta - p\circ T^i - \displaystyle\sum_{q\in\mathbb{P}}\lambda(q)(w(q)-q)\\[2ex] \qquad\qquad + \langle\mu,r^i\rangle = \sigma\\[1ex] \lambda\in\mathcal{F}\left(\mathbb{P},\mathbb{R}_+\right),\ \sigma\in\Sigma[x],\ \mu\in\Sigma[x]^{n_i},\\ \eta\in\mathbb{R},\\ \deg(\sigma)\leq 2m\deg T^i,\\ \deg(\langle\mu,r^i\rangle)\leq 2m\deg T^i\,. \end{array}\right. \qquad(6.3)$$

We conclude that, for all $i\in\mathcal{I}$, the evaluation of $F_i^{\mathcal{R}}$ can be done using SOS programming, since it is reduced to solve a minimization problem with a linear objective function and linear combination of polynomials constrained to be sum-of-squares.

Example 6.3. *We still consider the running example defined in Example 5.23 and take the following templates basis:* $q_1 : x \mapsto x_1^2$, $q_2 : x \mapsto x_2^2$, *and a well-chosen polynomial p of degree 6. For the index of the partition $i = 1$. Recall that $T^1(x) = \begin{pmatrix} 0.687x_1 + 0.558x_2 - 0.0001x_1x_2 \\ -0.292x_1 + 0.773x_2 \end{pmatrix}$ and $X^1 = \{x \in \mathbb{R}^2 \mid -x_1^2 + 1 \leq 0\}$ and thus $r_1^1(x) = -x_1^2 + 1$. Let $w \in \mathcal{F}(\mathbb{P}, \overline{\mathbb{R}})$, then:*

$$\left(F_1^{\mathcal{R}}(w)\right)(q_1) =$$

$$\inf_{\lambda, \sigma, \mu, \eta} \eta$$

$$\text{s.t.} \begin{cases} \eta - (0.687x_1 + 0.558x_2 - 0.0001x_1x_2)^2 \\ -\lambda(q_1)(w(q_1) - x_1^2) - \lambda(q_2)(w(q_2) - x_2^2) \\ -\lambda(p)(w(p) - p(x)) + \mu(x)(1 - x_1^2) = \sigma(x) \\ \lambda \in \mathcal{F}(\mathbb{P}, \mathbb{R}_+), \ \sigma \in \Sigma[x], \ \mu \in \Sigma[x], \ \eta \in \mathbb{R}, \\ \deg(\sigma) \leq 6, \ \deg(\mu) \leq 6. \end{cases}$$

In practice, one cannot find any feasible solution of degree less than 6, thus we replace the degree constraint by the more restrictive one: $\deg(\sigma) \leq 6$, $\deg(\mu) \leq 6$.

The computation of $F^{\sharp}$ requires the approximation of $X^{Init^{\dagger}} := \sup\{p(x), x \in X^{Init}\}$. Since X^{Init} is a basic semialgebraic set and each template p is a polynomial, then the evaluation of $X^{Init^{\dagger}}$ boils down to solving a polynomial maximization problem. Next, we use SOS reinforcement described above to over-approximate $X^{Init^{\dagger}}$ with the set $X^{Init^{\mathcal{R}}}$, defined as follows:

$$X^{Init^{\mathcal{R}}}(p) :=$$

$$\inf \left\{ \eta \ \middle| \ \begin{array}{c} \eta - p + \langle \nu^{n_{in}}, r^{n_{in}} \rangle = \sigma_0, \\ \eta \in \mathbb{R}, \ \sigma_0 \in \Sigma[x], \nu^{in} \in \Sigma[x]^{n_{in}}, \\ \deg(\sigma_0) \leq 2m, \deg(\langle \nu^{n_{in}}, r^{n_{in}} \rangle) \leq 2m \end{array} \right\}.$$

Thus, the value of $X^{Init^{\mathcal{R}}}(p)$ is obtained by solving an SOS optimization problem. Since X^{Init} is a nonempty compact basic semialgebraic set, this problem has a feasible solution (see the proof of [39, Th. 4.2]), ensuring that $X^{Init^{\mathcal{R}}}(p)$ is finite valued.

Example 6.4. *The initialization set X^{Init} of Example 5.23 is $[-1, 1] \times [-1, 1]$. It can be written as:* $\{(x_1, x_2) \in \mathbb{R}^2 \mid x_1^2 - 1 \leq 0, \ x_2^2 - 1 \leq 0\}$. *Then, considering the same template basis of Example 6.3 and the template q_1:*

$$X^{Init^{\mathcal{R}}}(q_1) :=$$

$$\inf \left\{ \eta \; \middle| \; \begin{array}{l} \eta - x_1^2 + \nu_1^{n_{in}}(x)(x_1^2 - 1) \\[4pt] + \nu_2^{n_{in}}(x)(x_2^2 - 1) = \sigma_0(x), \\[4pt] \eta \in \mathbb{R}, \; \sigma_0 \in \Sigma[x], \nu_1^{in}, \nu_2^{in} \in \Sigma[x], \\[4pt] \deg(\sigma_0) \le 6, \deg(\langle \nu_1^{n_{in}} \rangle) \le 6, \\[4pt] \deg(\langle \nu_2^{n_{in}} \rangle) \le 6 \end{array} \right\}.$$

*It is easy to see that taking for all $x \in \mathbb{R}^2$, $\nu_1^{n_{in}}(x) = 1$ and for all $x \in \mathbb{R}^2$, $\nu_2^{n_{in}}(x) = 0$
leads to $\eta - x_1^2 + \nu_1^{n_{in}}(x)(x_1^2 - 1) + \nu_2^{n_{in}}(x)(x_2^2 - 1) = \eta - 1 = \sigma_0(x)$. Thus for $\eta = 1$
and for all $x \in \mathbb{R}^2$, $\sigma_0(x) = 0$, we obtain $X^{Init^{\mathcal{R}}}(q_1) = 1$.*

Finally, we define the relaxed functional $F^{\mathcal{R}}$ for all $w \in \mathcal{F}(\mathbb{P}, \overline{\mathbb{R}})$ and for all
$p \in \mathbb{P}$ as follows:

$$\left(F^{\mathcal{R}}(w) \right)(p) = \sup \left\{ \sup_{i \in \mathcal{I}} \left(F_i^{\mathcal{R}}(w) \right)(p), X^{Init^{\mathcal{R}}}(p) \right\}. \tag{6.4}$$

By construction, the relaxed functional $F^{\mathcal{R}}$ provides a safe over-approximation
of the abstract semantics $F^{\sharp}$.

Proposition 6.5 (Safety).   *The following statements hold:*

1. *$X^{Init^{\dagger}} \le_{\mathbb{F}} X^{Init^{\mathcal{R}}}$;*
2. *for all $i \in \mathcal{I}$, for all $w \in \mathcal{F}(\mathbb{P}, \overline{\mathbb{R}})$, $F_i^{\sharp}(w) \le_{\mathbb{F}} F_i^{\mathcal{R}}(w)$;*
3. *for all $w \in \mathcal{F}(\mathbb{P}, \overline{\mathbb{R}})$, $F^{\sharp}(w) \le_{\mathbb{F}} F^{\mathcal{R}}(w)$.*

An important property that we will use to prove some results on policy
iteration algorithm is the monotonicity of the relaxed functional.

Proposition 6.6 (Monotonicity).

1. *For all $i \in \mathcal{I}$, $w \mapsto F_i^{\mathcal{R}}(w)$ is monotone on $\mathcal{F}(\mathbb{P}, \overline{\mathbb{R}})$;*
2. *the function $w \mapsto F^{\mathcal{R}}(w)$ is monotone on $\mathcal{F}(\mathbb{P}, \overline{\mathbb{R}})$.*

From the third assertion of Prop. 6.5, if w satisfies $F^{\mathcal{R}}(w) \le_{\mathbb{F}} w$ then $F^{\sharp}(w)$
$\le_{\mathbb{F}} w$ and from Prop. 6.1, $w^{\star}$ is an inductive invariant and thus $\mathcal{R} \subseteq w^{\star}$. This
result is formulated as the following corollary.

Corollary 6.7 (Over-approximation).   *For all $w \in \mathcal{F}(\mathbb{P}, \overline{\mathbb{R}})$ such that $F^{\mathcal{R}}$
$(w) \le_{\mathbb{F}} w$ then $\mathcal{R} \subseteq w^{\star}$.*

6.3.2   Policy iteration in polynomial templates abstract domains

We are interested in computing the least fixpoint $\mathcal{R}^{\mathcal{R}}$ of $F^{\mathcal{R}}$, $\mathcal{R}^{\mathcal{R}}$ being an over-
approximation of $\mathcal{R}$ (least fixpoint of F). As for the definition of $\mathcal{R}$, it can be
reformulated using Tarski's theorem as the minimal postfixpoint:

$$\mathcal{R}^{\mathcal{R}} = \min\{w \in \mathcal{F}\left(\mathbb{P}, \overline{\mathbb{R}}\right) \,|\, F^{\mathcal{R}}(w) \leq_{\mathbb{F}} w\}.$$

The idea behind policy iteration is to over-approximate $\mathcal{R}^{\mathcal{R}}$ using successive iterations which are composed of

- the computation of polynomial template bounds using linear programming, and
- the determination of new policies using SOS programming,

until a fixpoint is reached. Policy iteration navigates in the set of postfixpoints of $F^{\mathcal{R}}$ and needs to start from a postfixpoint w^0 know a priori. It acts like a narrowing operator and can be interrupted at any time. For further information on policy iteration, the interested reader can consult [52, 68].

6.3.3   Policies

Policy iteration can be used to compute a fixpoint of a monotone self-map defined as an infimum of a family of affine monotone self-maps. We propose to design a policy iteration algorithm to compute a fixpoint of $F^{\mathcal{R}}$. In this subsection, we give the formal definition of policies in the context of polynomial templates and define the family of affine monotone self-maps. We do not apply the concept of policies on $F^{\mathcal{R}}$ but on the functions $F_i^{\mathcal{R}}$ exploiting the fact that for all $i \in \mathcal{I}$, $F_i^{\mathcal{R}}$ is the optimal value of a minimization problem.

Policy iteration needs a *selection property*, that is, when an element $w \in \mathcal{F}\left(\mathbb{P}, \overline{\mathbb{R}}\right)$ is given, there exists a policy which achieves the infimum. In our context, since we apply the concept of policies to $F_i^{\mathcal{R}}$, it means that the minimization problem involved in the computation of $F_i^{\mathcal{R}}$ has an optimal solution. In our case, for $w \in \mathcal{F}\left(\mathbb{P}, \overline{\mathbb{R}}\right)$ and $p \in \mathbb{P}$, an optimal solution is a vector $(\lambda, \sigma, \mu) \in \mathcal{F}\left(\mathbb{P}, \mathbb{R}_+\right) \times \Sigma[x] \times \Sigma[x]^{n_i}$ such that, using (6.3), we obtain:

$$
\begin{aligned}
\left(F_i^{\mathcal{R}}(w)\right)(p) =& \\
p \circ T^i + \sum_{q \in \mathbb{P}} & \lambda(q)(w(q) - q) - \langle \mu, r^i \rangle + \sigma \\
\text{and } \deg(\sigma) \leq& \, 2m \deg T^i, \\
\deg(\langle \mu, r^i \rangle) \leq& \, 2m \deg T^i.
\end{aligned}
\tag{6.5}
$$

Observe that in Eq. (6.5), $\left(F_i^{\mathcal{R}}(w)\right)(p)$ is a scalar whereas the right-hand side is a polynomial. The equality in this equation means that this polynomial is a constant polynomial. Then we introduce the set of feasible solutions for the SOS problem $\left(F_i^{\mathcal{R}}(w)\right)(p)$:

$$
\mathrm{Sol}(w, i, p) = \left\{ \begin{array}{l} (\lambda, \sigma, \mu) \in \mathcal{F}\left(\mathbb{P}, \mathbb{R}_+\right) \times \Sigma[x] \times \Sigma[x]^{n_i} \\ \text{s.t. Eq. (6.5) holds} \end{array} \right\}.
\tag{6.6}
$$

Since policy iteration algorithm can be stopped at any step and still provides a sound over-approximation, we stop the iteration when $\mathrm{Sol}(w, i, p) = \emptyset$. Now, we are interested in the elements $w \in \mathcal{F}(\mathbb{P}, \mathbb{R})$ such that $\mathrm{Sol}(w, i, p)$ is nonempty:

$$\mathcal{FS}(\mathbb{P}, \overline{\mathbb{R}}) = \left\{ w \in \mathcal{F}(\mathbb{P}, \overline{\mathbb{R}}) \;\middle|\; \begin{array}{l} \forall i \in \mathcal{I},\ \forall p \in \mathbb{P}, \\ \mathrm{Sol}(w, i, p) \neq \emptyset \end{array} \right\}. \tag{6.7}$$

The notation $\mathcal{FS}(\mathbb{P}, \overline{\mathbb{R}})$ was introduced in [91] to define the elements $w \in \mathcal{F}(\mathbb{P}, \overline{\mathbb{R}})$ satisfying $\mathrm{Sol}(w, i, p) \neq \emptyset$. In [91, Section 4.3], $\mathrm{Sol}(w, i, p) \neq \emptyset$ using Slater's constraint qualification condition. In the current nonlinear setting, we cannot use the same condition, which yields a more complicated definition for $\mathcal{FS}(\mathbb{P}, \overline{\mathbb{R}})$.

Finally, we can define a policy as a map which selects, for all $w \in \mathcal{FS}(\mathbb{P}, \overline{\mathbb{R}})$, for all $i \in \mathcal{I}$, and for all $p \in \mathbb{P}$, a vector of $\mathrm{Sol}(w, i, p)$. More formally, we have the following definition:

Definition 6.8 (Policies in the SOS Policy Iteration). *A policy is a map* $\pi : \mathcal{FS}(\mathbb{P}, \overline{\mathbb{R}}) \mapsto ((\mathcal{I} \times \mathbb{P}) \mapsto \mathcal{F}(\mathbb{P}, \mathbb{R}_+) \times \Sigma[x] \times \Sigma[x]^{n_i} \times \Sigma[x]^{n_o})$ *such that:* $\forall w \in \mathcal{FS}(\mathbb{P}, \overline{\mathbb{R}}),\ \forall i \in \mathcal{I},\ \forall p \in \mathbb{P},\ \pi(w)(i, p) \in \mathrm{Sol}(w, i, p)$.

We denote by Π the set of policies. For $\pi \in \Pi$, let us define π_λ as the map from $\mathcal{FS}(\mathbb{P}, \overline{\mathbb{R}})$ to $(\mathcal{I} \times \mathbb{P}) \mapsto \mathcal{F}(\mathbb{P}, \mathbb{R}_+)$ which associates with $w \in \mathcal{FS}(\mathbb{P}, \overline{\mathbb{R}})$ and $(i, p) \in \mathcal{I} \times \mathbb{P}$ the first tuple element of $\pi(w)(i, p)$, i.e., if $\pi(w)(i, p) = (\lambda, \sigma, \mu)$ then $\pi_\lambda(w)(i, p) = \lambda$.

As said before, policy iteration exploits the linearity of maps when a policy is fixed. We have to define the affine maps we will use in a policy iteration step. With $\pi \in \Pi$, $w \in \mathcal{FS}(\mathbb{P}, \overline{\mathbb{R}})$, $i \in \mathcal{I}$ and $p \in \mathbb{P}$ and $\lambda = \pi_\lambda(w)(i, p)$, let us define the map $\phi_{w,i,p}^\lambda : \mathcal{F}(\mathbb{P}, \overline{\mathbb{R}}) \mapsto \overline{\mathbb{R}}$ as follows:

$$v \mapsto \phi_{w,i,p}^\lambda(v) = \sum_{q \in \mathbb{P}} \lambda(q) v(q) + \left(F_i^{\mathcal{R}}(w)\right)(p) - \sum_{q \in \mathbb{P}} \lambda(q) w(q). \tag{6.8}$$

Then, for $\pi \in \Pi$, we define for all $w \in \mathcal{FS}(\mathbb{P}, \overline{\mathbb{R}})$, the map $\Phi_w^{\pi(w)}$ from $\mathcal{F}(\mathbb{P}, \overline{\mathbb{R}}) \mapsto \mathcal{F}(\mathbb{P}, \overline{\mathbb{R}})$. Let $v \in \mathcal{F}(\mathbb{P}, \overline{\mathbb{R}})$ and $p \in \mathbb{P}$:

$$\Phi_w^{\pi(w)}(v)(p) = \sup \left\{ \sup_{i \in \mathcal{I}} \phi_{w,i,p}^\lambda(v),\ X^{Init\,\mathcal{R}}(p) \right\}. \tag{6.9}$$

Example 6.9. *Let us consider Example 6.3 and the function* $w^0(q_1) = w^0(q_2) = 2.1391$ *and* $w^0(p) = 0$. *Then there exists two SOS polynomials* $\overline{\mu}$ *and* $\overline{\sigma}$ *such that, for all* $x \in \mathbb{R}^d$:
$$\left(F_1^{\mathcal{R}}(w)\right)(q_1) = (0.687x_1 + 0.558x_2 - 0.0001x_1x_2)^2 + \overline{\lambda}(q_1)(2.1391 - x_1^2) + \overline{\lambda}$$
$(q_2)(2.1391 - x_2^2) - \overline{\lambda}(p)p(x) - \overline{\mu}(x)(1 - x_1^2) + \overline{\sigma}(x) = 1.5503$ *with* $\overline{\lambda}(q_1) = \overline{\lambda}(q_2)$ $= 0$ *and* $\overline{\lambda}(p) = 2.0331$. *It means that* $\overline{\lambda}, \overline{\mu}$ *and* $\overline{\sigma}$ *are computed such that* $(0.687x_1$

$+0.558x_2 - 0.0001x_1x_2)^2 + \overline{\lambda}(q_1)(2.1391 - x_1^2) + \overline{\lambda}(q_2)(2.1391 - x_2^2) - \overline{\lambda}(p)p(x)$
$-\overline{\mu}(x)(1 - x_1^2) + \overline{\sigma}(x)$ *is actually a constant polynomial.*

Then $(\overline{\lambda}, \overline{\mu}, \overline{\sigma}) \in \mathrm{Sol}(w^0, 1, q_1)$ *and we can define a policy* $\pi(w^0)$ *such that* $\pi(w^0)(1, q_1) = (\overline{\lambda}, \overline{\mu}, \overline{\sigma})$ *and thus* $\pi_\lambda(w^0)(1, q_1) = (0, 0, 2.0331)$. *We can thus define for* $v \in \mathcal{F}(\mathbb{P}, \mathbb{R})$, *the affine mapping:* $\phi^\lambda_{w^0, 1, q_1}(v) = \lambda(q_1)v(q_1) + \lambda(q_2)v(q_2) + \lambda(p)v$ $(p) + \left(F_1^{\mathcal{R}}(w)\right)(q_1) - \lambda(q_1)w(q_1) - \lambda(q_2)w(q_2) - \lambda(p)w(p) = 2.1391v(p) + 1.5503.$

Let us denote by $\mathcal{F}(\mathbb{P}, \mathbb{R})$ the set of finite valued function on $\mathbb{P}$, i.e., $g \in \mathcal{F}(\mathbb{P}, \mathbb{R})$ iff $g(p) \in \mathbb{R}$ for all $p \in \mathbb{P}$.

Proposition 6.10 (Properties of $\phi^\lambda_{i, w, p}$). *Let* $\pi \in \Pi$, $w \in \mathcal{FS}(\mathbb{P}, \overline{\mathbb{R}})$, *and* $(i, p) \in \mathcal{I} \times \mathbb{P}$. *Let us write* $\lambda = \pi_\lambda(w)(i, p)$. *The following properties are true:*

1. $\phi^\lambda_{w, i, p}$ *is affine on* $\mathcal{F}(\mathbb{P}, \mathbb{R})$;
2. $\phi^\lambda_{w, i, p}$ *is monotone on* $\mathcal{F}(\mathbb{P}, \overline{\mathbb{R}})$;
3. $\forall v \in \mathcal{F}(\mathbb{P}, \overline{\mathbb{R}})$, $F_i^{\mathcal{R}}(v)(p) \le \phi^\lambda_{w, i, p}(v)$;
4. $\phi^\lambda_{w, i, p}(w) = F_i^{\mathcal{R}}(w)(p)$.

The properties presented in Prop. 6.10 imply some useful properties for the maps $\Phi_w^{\pi(w)}$.

Proposition 6.11 (Properties of $\Phi_w^{\pi(w)}$). *Let* $\pi \in \Pi$ *and* $w \in \mathcal{FS}(\mathbb{P}, \overline{\mathbb{R}})$. *The following properties are true:*

1. $\Phi_w^{\pi(w)}$ *is monotone on* $\mathcal{F}(\mathbb{P}, \overline{\mathbb{R}})$;
2. $F^{\mathcal{R}} \le_{\mathbb{F}} \Phi_w^{\pi(w)}$;
3. $\Phi_w^{\pi(w)}(w) = F^{\mathcal{R}}(w)$;
4. *Suppose that the least fixpoint of* $\Phi_w^{\pi(w)}$ *is* $L \in \mathcal{F}(\mathbb{P}, \mathbb{R})$. *Then* L *can be computed as the unique optimal solution of the linear program:*

$$\inf \left\{ \sum_{p' \in \mathbb{P}} v(p') \left| \begin{array}{l} \forall (i, p) \in \mathcal{I} \times \mathbb{P}, \\[2mm] \phi^{\pi_\lambda(w)(i, p)}_{i, w, p}(v) \le v(p), \\[2mm] \forall q \in \mathbb{P}, \\[2mm] X^{Init \mathcal{R}}(q) \le v(q) \end{array} \right. \right\}. \tag{6.10}$$

Recall that a function $g : \mathbb{R}^d \mapsto \mathbb{R}$ is upper-semicontinuous at x iff for all $(x_n)_{n \in \mathbb{N}}$ converging to x, then $\limsup_{n \to +\infty} g(x_n) \le g(x)$.

Proposition 6.12. *Let* $p \in \mathbb{P}$. *Then* $w \mapsto F^{\mathcal{R}}(w)(p)$ *is upper-semicontinuous on* $\mathcal{FS}(\mathbb{P}, \overline{\mathbb{R}}) \cap \mathcal{F}(\mathbb{P}, \mathbb{R})$.

input : $w^0 \in \mathcal{F}(\mathbb{P}, \mathbb{R})$, a postfixpoint of $F^{\mathcal{R}}$
output: a fixpoint $w = F^{\mathcal{R}}(w)$ if $\forall k \in \mathbb{N}$, $w^k \in \mathcal{FS}(\mathbb{P}, \overline{\mathbb{R}})$ or a postfixpoint
       otherwise

1 k=0;
2 **while** *fixpoint not reached* **do**
3 | **begin** compute the next policy π for the current iterate w^k
4 | | Compute $F^{\mathcal{R}}(w^k)$ using Eq. (6.4) and Eq. (6.3);
5 | | **if** $w^k \in \mathcal{FS}(\mathbb{P}, \overline{\mathbb{R}})$ **then**
6 | | | Define $\pi(w^k)$;
7 | | **else**
8 | | | return w^k;
9 | | **end**
10 | **end**
11 | **begin** compute the next iterate w^{k+1}
12 | | Define $\Phi_{w^k}^{\pi(w^k)}$ and compute the least fixpoint w^{k+1} of $\Phi_{w^k}^{\pi(w^k)}$ from
       | | Problem (6.10);
13 | | k=k+1;
14 | **end**
15 **end**

Figure 6.1. SOS-based policy iteration algorithm for PPS programs.

6.3.4   Policy iteration

Now, we describe the policy iteration algorithm. We suppose that we have a postfixpoint w^0 of $F^{\mathcal{R}}$ in $\mathcal{F}(\mathbb{P}, \mathbb{R})$.

Now let us detail, step by step, the algorithm presented in Figure 6.1. At Line 1, the algorithm is initialized and thus $k = 0$. At Line 4, we compute $F^{\mathcal{R}}(w^k)$ using Eq. (6.4) and solve the SOS problem involved in Eq. (6.3). At Line 6, if for all $i \in \mathcal{I}$ and for $p \in \mathbb{P}$, the SOS problem involved in Eq. (6.3) has an optimal solution, then a policy π is available and we can choose any optimal solution of the SOS problem involved in Eq. (6.3) as policy. If an optimal solution does not exist then the algorithm stops and returns w^k. Now, if a policy π has been defined, the algorithm goes to Line 12 and we can define $\Phi_{w^k}^{\pi(w^k)}$ following Eq. (6.9). Then, we solve LP problem (6.10) and define the new bound on templates w^{k+1} as the smallest fixpoint of $\Phi_{w^k}^{\pi(w^k)}$. Finally, at Line 13, k is incremented.

If for some $k \in \mathbb{N}$, $w^k \notin \mathcal{FS}(\mathbb{P}, \overline{\mathbb{R}})$ and $w^{k-1} \in \mathcal{FS}(\mathbb{P}, \overline{\mathbb{R}})$ then the algorithm stops and returns w^k. Hence, we set for all $l \geq k$, $w^l = w^k$.

Theorem 6.13 (Convergence result of Algorithm 6.1). *The following statements hold:*

1. *For all $k \in \mathbb{N}$, $w^k \in \mathcal{F}(\mathbb{P}, \mathbb{R})$ and $F^{\mathcal{R}}(w^k) \leq w^k$.*
2. *The sequence $(w^k)_{k \geq 0}$ generated by the algorithm is decreasing and converges.*
3. *Let $w^{\infty} = \lim_{k \to +\infty} w^k$, then $F^{\mathcal{R}}(w^{\infty}) \leq w^{\infty}$. Furthermore, if for all $k \in \mathbb{N}$, $w^k \in \mathcal{FS}(\mathbb{P}, \overline{\mathbb{R}})$ and if $w^{\infty} \in \mathcal{FS}(\mathbb{P}, \overline{\mathbb{R}})$ then $F^{\mathcal{R}}(w^{\infty}) = w^{\infty}$.*

6.4   EXAMPLE

Recall that our running example is given by the following piecewise polynomial system: $(X^{Init}, \{X^1, X^2\}, \{T^1, T^2\})$, where:

$$X^{Init} = [-1, 1] \times [-1, 1]$$
$$\begin{cases} X^1 = \{x \in \mathbb{R}^2 \mid -x_1^2 + 1 \le 0\} \\ X^2 = \{x \in \mathbb{R}^2 \mid x_1^2 - 1 < 0\} \end{cases}$$

and the functions relative to the partition $\{X^1, X^2\}$ are:

$$T^1(x_1, x_2) = \begin{pmatrix} 0.687x_1 + 0.558x_2 - 0.0001x_1x_2 \\ -0.292x_1 + 0.773x_2 \end{pmatrix}$$

and

$$T^2(x_1, x_2) = \begin{pmatrix} 0.369x_1 + 0.532x_2 - 0.0001x_1^2 \\ -1.27x_1 + 0.12x_2 - 0.0001x_1x_2 \end{pmatrix}.$$

The first step consists in constructing the template basis and computing the template p and bound w on the reachable values as a solution of Problem (5.56). We fix the degree of p to 6. The template p generated from MATLAB is of degree 6 and is defined as follows:

$$-1.931348006 + 3.5771x_1^2 + 2.0669x_2^2 + 0.7702x_1x_2 - (2.6284\text{e–}4)x_1^3 -$$
$$(5.5572\text{e–}4)x_1^2x_2 + (3.1872\text{e–}4)x_1x_2^2 + 0.0010x_2^3 - 2.4650x_1^4 - 0.5073x_1^3x_2 -$$
$$2.8032x_1^2x_2^2 - 0.5894x_1x_2^3 - 1.4968x_2^4 + (2.7178\text{e–}4)x_1^5 + (1.2726\text{e–}4)x_1^4x_2 -$$
$$(3.8372\text{e–}4)x_1^3x_2^2 + (6.5349\text{e–}5)x_1^2x_2^3 + (5.7948\text{e–}6)x_1x_2^4 - (6.2558\text{e–}4)x_2^5 +$$
$$0.5987x_1^6 - 0.0168x_1^5x_2 + 1.1066x_1^4x_2^2 + 0.3172x_1^3x_2^3 + 0.8380x_1^2x_2^4 +$$
$$0.0635x_1x_2^5 + 0.4719x_2^6.$$

The upper bound w is equal to 2.1343. In order to compute bounds per variable, we can take the template basis $\mathbb{P} = \{p, x \mapsto x_1^2, x \mapsto x_2^2\}$. We write q_1 for $x \mapsto x_1^2$ and q_2 for $x \mapsto x_2^2$. The basic semialgebraic $\{x \in \mathbb{R}^2 \mid p(x) \le 0,\ q_1(x) \le 2.1343,\ q_2(x) \le 2.1343\}$ is an inductive invariant and the corresponding bounds function is $w^0 = (w^0(q_1), w^0(q_2), w^0(p)) = (2.1343, 2.1343, 0)$.

As in Line 4 of Algorithm 6.1, we compute the image of w^0 by $F^{\mathcal{R}}$ using SOS (Eq. (6.3)). We found that

$$F^{\mathcal{R}}(w^0)(q_1) = 1.5503,$$
$$F^{\mathcal{R}}(w^0)(q_2) = 1.9501,$$
$$F^{\mathcal{R}}(w^0)(p) = 0.$$

Since $w^0 \in \mathcal{FS}\left(\mathbb{P}, \overline{\mathbb{R}}\right)$, Algorithm 6.1 goes to Line 6 and the computation of $F^{\mathcal{R}}(w^0)$ permits us to determine a new policy $\pi(w^0)$. The important data is the

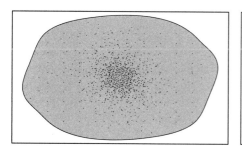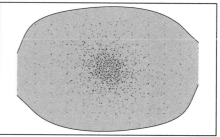

Figure 6.2. Successive sets computed from policy iterations. In gray, on the left, the set $w^{0^\star}$ computed from Problem (5.56), while, on the right, the set $w^{1^\star}$ computed from policy iterations. In both figures, black points represent a discretized version of $\mathcal{R}$.

vector λ. For example, for $i = 1$ and the template q_1, the vector λ is $(0, 0, 2.0331)$. It means that we associate for each template q a weight $\lambda(q)$. In the case of $\lambda = (0, 0, 2.0331)$, $\lambda(q_1) = 0$, $\lambda(q_2) = 0$, and $\lambda(p) = 2.0332$. For $i = 1$, the template q_1 and the bound vector w^0, the function $\phi^\lambda_{w^0, 1, q_1}(v) = 2.0331v(p) + 1.5503$.

To get the new invariant, the algorithm goes to Line 12 and we compute a bound vector w^1 solution of Linear Program (6.10). In this case, it corresponds to the following LP problem:

$$\min v(q_1) + v(q_2) + v(p)$$

s.t.

$$\begin{cases} 1 \leq v(q_1), \ 1 \leq v(q_2), \ 0 \leq v(p) & \text{(init)} \\ 0.4578v(p) + 0.8843 \leq v(q_1), \\ 0.2048v(p) + 1.9501 \leq v(q_2), & (i = 1) \\ 0.9985v(p) - 3.4691\text{e-}7 \leq v(p) \\ 2.0331v(p) + 1.5503 \leq v(q_1), \\ 1.0429v(p) + 1.2235 \leq v(q_2), & (i = 2) \\ 0.9535v(p) - 0.0248 \leq v(p). \end{cases}$$

We obtain:

$$w^1(q_1) = 1.5503, \ w^1(q_2) = 1.9501 \text{ and } w^1(p) = 0.$$

We then come back to Line 4 of the algorithm and we compute $F^{\mathcal{R}}(w^1)$ using the SOS program Eq. (6.3). The implemented stopping rule is $\|F^{\mathcal{R}}(w^k) - w^k\|_\infty \leq$ 1e–6 and since $\|F^{\mathcal{R}}(w^1) - w^1\|_\infty \leq$ 1e–6, the algorithm terminates. The two successive inductive invariants are depicted in Figure 6.2.

6.5 RELATED WORKS

As presented in the Formal Methods introduction in chapter 2, the classical framework for abstract interpretation is a fixpoint over-approximation through a Kleene fixpoint computation using widening to ensure convergence. Another mechanism, narrowing, enables us to recover precision once a postfixpoint has been obtained through widening.

In static analysis, the more recent approach of policy[1] iterations [52, 68, 69] attempts to solve exactly the fixpoint equation for a given abstract domain when specific conditions are satisfied using appropriate mathematical solvers. While this chapter addressed a rather large set of programs–piecewise polynomial systems–using SOS optimization, related (and previous) works were considering simpler classes of programs and of convex optimizations. For example, when both the abstract domain and the fixpoint equation use linear equations, then linear programming could be used to compute the exact solution without the need of widening and narrowing [68, 69]. Similarly, when the function and the abstract domain are at most quadratic, semidefinite programming (SDP) could be used [91, 93, 110]. In all cases, these analyses are performed on template-based abstract domains, representing the abstract elements as sublevel sets; optimization techniques are used to bound these templates.

Regarding policy iterations' related works, two different "schools" exist in the static analysis community. The "French school" [52, 68, 91, 110] offers to iterate on min-policies, starting from an over-approximation of a fixpoint and decreasing the bounds until the fixpoint is reached. The "German school" [69, 70, 110], in contrast, operates on max-policies, starting from the bottom and increasing the bounds until a fixpoint is reached. While the first can be interrupted at any point leaving a sound over-approximation, the second approach requires waiting until the fixpoint is reached to provide its result. Note that a first valid postfixpoint is required in the first case.

6.5.1 Min-policy iterations

To some extent, min-policy iterations [91] can be seen as a very efficient *narrowing*, since they perform descending iterations from a postfixpoint towards some fixpoint, working in a way similar to the Newton-Raphson method. Iterations are not guaranteed to reach a fixpoint but can be stopped at any time leaving an over-approximation thereof. Moreover, convergence is usually fast.

Writing a system of equations $b = F(b)$ with $b = (b_i)_{i \in [\![1,n]\!]}$ and $F : \overline{\mathbb{R}}^n \to \overline{\mathbb{R}}^n$ (n being the number of templates), a min-policy is defined as follows: $\underline{F}$ is a min-policy for F if for every $b \in \overline{\mathbb{R}}^n$, $F(b) \leq \underline{F}(b)$ and there exist some $b_0 \in \overline{\mathbb{R}}^n$ such that $\underline{F}(b_0) = F(b_0)$.

The following theorem can then be used to compute the least fixpoint of F.

---

[1] The word *strategy* is also used in the literature for *policy*, with equivalent meaning.

Theorem 6.14. *Given a (potentially infinite) set $\underline{\mathcal{F}}$ of min-policies for F. If for all $b \in \overline{\mathbb{R}}^n$ there exist a policy $\underline{F} \in \underline{\mathcal{F}}$ interpolating F at point b (i.e., $\underline{F}(b) = F(b)$) and if each $\underline{F} \in \underline{\mathcal{F}}$ has a least fixpoint $\mathrm{lfp}\underline{F}$, then the least fixpoint of F satisfies*

$$\mathrm{lfp}F = \bigwedge_{\underline{F} \in \underline{\mathcal{F}}} \mathrm{lfp}\underline{F}.$$

Iterations are done with two main objects: a min-policy σ and a tuple β of values for variables b_i of the system of equations. The following policy iteration algorithm starts from some postfixpoint β_0 of F and aims at refining it to produce a better over-approximation of a fixpoint of F. Policy iteration algorithms always proceed by iterating two phases: first a policy σ_i is selected, then it is solved giving some β_i. More precisely in our case:

- find a linear min-policy σ_{i+1} being tangent to F at point β_i (this can be done thanks to a semidefinite programming solver and an appropriate relaxation);
- compute the least fixpoint β_{i+1} of policy σ_{i+1} thanks to a linear programming solver.

Iterations can be stopped at any point (for instance, after a fixed number of iterations or when progress between β_i and β_{i+1} is considered small enough), leaving an over-approximation β of a fixpoint of F.

6.5.2 Max-policy iterations

Behaving somewhat as a super *widening*, max-policy iterations [93] work in the opposite direction compared to min-policy iterations. They start from the bottom and iterate computations of greatest fixpoints on so-called max-policies until a global fixpoint is reached. Unlike the previous approach, the algorithm terminates with a *theoretically* precise fixpoint, but the user has to wait until the end since intermediate results are not over-approximations of a fixpoint. Max-policies are the dual of min-policies: $\overline{F}$ is a max-policy for F if for every $b \in \overline{\mathbb{R}}^n$, $\overline{F} \leq F(b)$ and there exist some $b_0 \in \overline{\mathbb{R}}^n$ such that $\overline{F}(b_0) = F(b_0)$. In particular, the choice of one term in each equation is a max-policy.

Iterations are done with two main objects: a max-policy σ and a tuple β of values for variables $b_{i,j}$ of the system of equations. Considering that computing a fixpoint on a given policy reduces to a mathematical optimization problem and that a fixpoint of the whole equation system is also a fixpoint of some policy, the following policy iteration algorithm aims at finding such a policy by solving optimization problems. To initiate the algorithm, a term $-\infty$ is added to each equation, the initial policy σ_0 is then $-\infty$ for each equation, and the initial value β_0 is the tuple $(-\infty, \ldots, -\infty)$. Then policies are iterated:

- find a policy σ_{i+1} improving policy σ_i at point β_i, i.e., that reaches (strictly) greater values evaluated at point β_i; if none is found, exit;
- compute the greatest fixpoint β_{i+1} of policy σ_{i+1}.

The last tuple β is then a fixpoint of the whole system of equations. The max-policy iteration builds an ascending chain of abstract elements similar to Kleene iterations elements. However, it is guaranteed to be finite, bounded by the number of policies σ, while Kleene iterations require the use of widening to ensure termination. Since there are exponentially many max-policies in the number of templates and since each policy can be an improving one only once, we have an exponential bound on the number of iterations. But in practice, only a small number of policies are usually considered and the number of iterations remains reasonable. One approach to selecting a good policy is to rely on SMT solvers to find a matching policy [137].

Last, recent works [146] relied on max-policies based on linear problems and linear programming solvers to compute efficiently local invariants on large programs. This work is applied in a completely different context than ours: targeting general C programs rather than critical controllers, with linear properties rather than expressive semialgebraic ones. It shows the applicability of the approach to a larger set of programs than numerical controllers.

Part III

System-level Analysis at
Model and Code Level

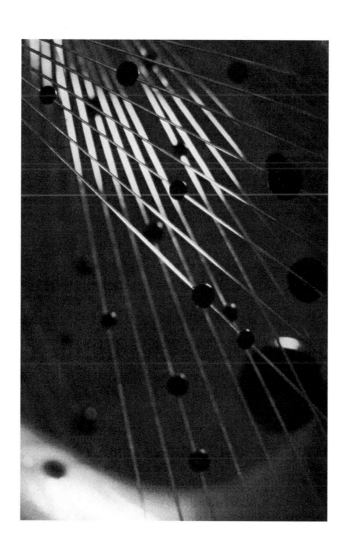

Chapter Seven

System-level Properties as Numerical Invariants

ALL NUMERICAL TOOLS presented in previous chapters were focused on the precise over-approximation of reachable states. In terms of properties addressed, we could argue about simple properties: e.g., the state space is bounded, the reachable states avoid a bad region, etc.

However, we believe that it is important to be able to express higher level properties than just bounding reachable states.

The idea that drove our invariants and template synthesis was this notion of Lyapunov functions and of Lyapunov stability. Assuming a control level property, it would be extremely interesting to be able to express this property over the code or model artifact.

A main limitation for the study of these control level properties is the need for the plant description, which is generally not available when considering code artifact. In the following we assume the plant semantics is provided in a discrete fashion and therefore amenable to code level description as presented in chapter 3.

We summarize here an attempt to express classical notions of control theory such as stability or robustness using the previously presented invariant-based tools.

NOTATIONS

Let us first recall the notations of chapter 3, we focus on linear systems, i.e., a linear plant with a linear controller feedback. Both the controller and the plant dynamics are expressed as discrete linear systems. Let (A^c, B^c, C^c, D^c) and (A^p, B^p, C^p) the matrices defining the controller and plant dynamics, respectively; e denotes the input of the controller, often referred to as the error, i.e., the distance to the target reference in; u denotes both the output of the controller and the input of the plant, such as the effect of actuator commands; y denotes the measure of the plant state, i.e., the feedback, e.g., as obtained by sensors:

$$\begin{cases} x^c_{k+1} &= A^c x^c_k + B^c e_k \\ u_k &= C^c x^c_k + D^c e_k \end{cases} \qquad \begin{cases} x^p_{k+1} &= A^p x^p_k + B^p u_k \\ y_k &= C^p x^p_k \end{cases} \qquad (7.1)$$

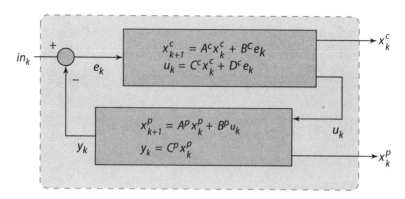

Figure 7.1. Closed-loop system.

A closed-loop representation of the system is given in Fig. 7.1; it is expressed over the state space defined by vectors $\begin{pmatrix} x^c & x^p \end{pmatrix}^\mathsf{T}$. Let $\mathbf{x}$ be such vectors. The error e is computed using a reference command in and the feedback y obtained from the plant.

$$e_k = in_k - y_k$$

One can consider in as the input of the closed-loop system, and $\mathbf{x}$ as its output.

7.1   OPEN-LOOP AND CLOSED-LOOP STABILITY

As mentioned in chapter 3, the notion of stability for a dynamical system captures both the boundedness of reachable states and a notion of convergence. A stable system guarantees that a small change in the input will not produce a large change in the output. Mathematically speaking, the notion of asymptotic stability ensures that with a null input, the system converges to zero. This stability can be studied in two ways: open-loop stability and closed-loop stability. In the open-loop setting the stability of the controller itself is studied while in the closed-loop setting the complete system integrating the feedback interconnection of controller and plant is addressed.

While closed-loop stability is the main stability property of interest–that is, the controlled system will have a stable behavior–ensuring open-loop stability avoids the undesirable situation where the feedback interconnection is stable, while the controller alone is intrinsically unstable. In terms of system implementation, an open-loop stable controller has a reasonable behavior on its own, e.g., assuming only bounded input, it will provide a bounded output. This is called the bounded input bounded output (BIBO) property.

Stability properties can be assessed in different ways. A system's dynamics are expressed as transfer functions mapping inputs to outputs. These are obtained by taking the Fourier or Laplace transform of the impulse response of a system. This so-called frequency domain approach is commonly used for linear systems, along with graphical tools such as Bode plots or Nyquist diagrams. An alternative approach, temporal domain analysis, is performed on the state-space representation, and is based on Lyapunov functions. As mentioned in the previous parts, Lyapunov functions express a notion of positive energy that decreases along the trajectories of the system and captures its asymptotic stability. For linear systems, such functions are usually defined using a positive definite matrix $P \succeq 0$ such that:

$$A^{\mathsf{T}} P A - P \prec 0, \tag{7.2}$$

where A is the state matrix of the system.

7.1.1 Lyapunov function computation

Using the tools proposed in chapter 5 we can compute inductive numerical invariants, such as positive definite matrices P^o and P^c denoting Lyapunov functions for these open- and closed-loop systems.

For the open-loop system, the Lyapunov function P^o is used to express a BIBO property of the controller alone: to bound reachable states x^c assuming a bounded input e:

$$\|e\|_\infty \le 1 \implies x^{c\mathsf{T}} P^o x^c \le 1.$$

For the closed-loop system, integrating the feedback of the plant in the controller input, a similar property is expressed. For a bounded target reference in, the closed-loop system will admit only bounded reachable states $\mathbf{x}$:

$$\|in\|_\infty \le 1 \implies \mathbf{x}^{\mathsf{T}} P^c \mathbf{x} \le 1.$$

These boundedness properties may seem weak to control engineers compared to the asymptotic stability properties expressed by the Lyapunov functions. However, they are of extreme importance to guarantee that the implementation will behave properly, without diverging and causing runtime errors, e.g., producing numerical overflows. Once provided with a quadratic bound on reachable states using the Lyapunov function characterizing the stability of the controller, static analyses of the discrete model and the code can rely on policy iterations, cf. chap. 6, to infer bounds on x^c and $\mathbf{x}$.

7.1.2 Stability of closed-loop system without saturation

Recall that the closed-loop system example presented in chapter 3 was presented in two flavors. The first one considered a simple feedback between the linear controller and the linear plant. This global linear system is exactly described by Figure. 7.1.

```
xc1 = xc2 = xp1 = xp2 = 0;
while (1) {
  yd = acquire_input();
  assert(yd >= -0.5 && yd <= 0.5);
  oxc1 = xc1; oxc2 = xc2; oxp1 = xp1; oxp2 = xp2;
  xc1 = 0.499 * oxc1 - 0.05 * oxc2 + (oxp1 - yd);
  xc2 = 0.01 * oxc1 + oxc2;
  xp1 = 0.028224 * oxc1 + oxp1 + 0.01 * oxp2
        - 0.064 * (oxp1 - yd);
  xp2 = 5.6448 * oxc1 - 0.01 * oxp1 + oxp2
        - 12.8 * (oxp1 - yd);
  wait_next_clock_tick();
}
```

Figure 7.2. Analyzed code for the closed-loop system.

Figure 7.3. Control flow graph for code of Figure 7.2.

Figure 7.2 displays the analyzed code for the closed-loop system described in the previous section. From such a code, one extracts the control flow graph of Figure 7.3.

Our analysis then synthesizes a quadratic Lyapunov-based template P, inductive over system transitions:

$$A^\mathsf{T} P A - P \prec 0, \tag{7.3}$$

where A denotes the closed-loop system discrete dynamics.

Let P be the matrix obtained:

$$P := \begin{bmatrix} 1.7776 & 1.3967 & -0.6730 & 0.1399 \\ 1.3967 & 1.1163 & -0.4877 & 0.1099 \\ -0.6730 & -0.4877 & 0.3496 & -0.0529 \\ 0.1399 & 0.1099 & -0.0529 & 0.0111 \end{bmatrix}.$$

From the extracted control flow graph and a set of expressions t_i on program variables, called *templates*, policy iterations techniques, cf. chapter 6, compute, for each graph vertex, bounds b_i such that $\bigwedge_i t_i \leq b_i$ is an invariant.

Given the templates $t_1 := x^\mathsf{T} P x$, $t_2 := x_{c1}^2$, $t_3 := x_{c2}^2$, $t_4 := x_{p1}^2$, and $t_5 := x_{p2}^2$ where x is the vector $[x_{c1}\, x_{c2}\, x_{p1}\, x_{p2}]^\mathsf{T}$ and (rounded to four digits) policy iterations compute the invariant

$$t_1 \leq 0.2302 \wedge t_2 \leq 51.0162 \wedge t_3 \leq 15.4720$$
$$\wedge\, t_4 \leq 10.1973 \wedge t_5 \leq 1767.75$$

which implies

$$|x_{c1}| \leq 7.1426 \wedge |x_{c2}| \leq 3.9334 \wedge |x_{p1}| \leq 3.1933$$
$$\wedge\, |x_{p2}| \leq 42.0446.$$

Our static analyzer took 0.76s to produce this template and 1.28 to bound it on an Intel Core2 @ 1.2GHz, hence a fully automatic computation in a total of 2.19s. This shows the existence of a Lyapunov function, bounds reachable states, and proves stability.

7.1.3   Closed-loop system with saturations

Realistic controllers usually contain saturations to bound the values read from sensors or sent to actuators, in order to ensure that these values remain in the operating ranges of those devices. With such a saturation on its input, the control flow graph of our running example changes to the one shown in Figure 7.4. Unfortunately, the previous method does not readily apply for the system with a saturation (or saturations).

A first idea could be to try to generate, as previously described, a quadratic template P for each edge of the control flow graph of Figure 7.4. This approach sometimes proves successful but fails on our running example. Indeed, only one of the edges of the graph in Figure 7.4 leads to a template P (for other edges, the Lyapunov equation has no solution) and this template does not allow policy iterations to compute a worthwhile invariant on the whole program.

Using common Lyapunov functions constitutes a second idea, that is, looking for a solution to the conjunction of Lyapunov equations for each edge. Again, this fails since Lyapunov equations have no solution for some edges. This is due to the fact that the closed-loop system is not globally stable. Indeed, intuitively, when its input is saturated, the controller is not able to stabilize any arbitrary state of the plant.

Last, other approaches such as piecewise Lyapunov functions, cf. Sec. 5.3, admit similar limits: they require strict inequalities to ensure soundness of the analysis.

The following two Sections 7.1.3.1 and 7.1.3.2 offer two alternative ways to generate a template $x^\mathsf{T} P x$ such that $x^\mathsf{T} P x \leq r$ is an invariant of the closed-loop system with saturation for some r. Both methods manage to produce such a template but more investigations are needed to determine their relative advantages and drawbacks.

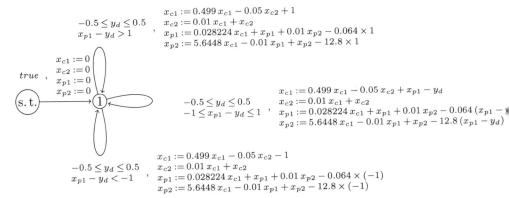

Figure 7.4. Control flow graph for the system with a saturation.

7.1.3.1 Linearizing the saturation

One solution in this case, strongly inspired from [92], provides a heuristic that can be used on systems with saturations, such as the one described in Equation (3.7). Indeed, let P be a candidate matrix describing an invariant ellipsoid for the system. We try to characterize P as closely as possible while keeping the solving process tractable.

Assuming $x_k^T P x_k \leq 1$, a bound on $|C x_k|$ is given by $\gamma := \sqrt{C P^{-1} C^T}$. Since $|y_{d,k}| \leq 0.5$, the constant $\tilde{\gamma} := \gamma + 0.5$ is an upper bound on $|C x_k - y_{d,k}|$. Letting $y_{c,k} := \text{SAT}(C x_k - y_{d,k})$, we have the following sector bound:

$$\left(y_{c,k} - \frac{1}{\tilde{\gamma}} (C x_k - y_{d,k}) \right) (y_{c,k} - (C x_k - y_{d,k})) \leq 0. \tag{7.4}$$

Figure 7.5 illustrates the reason for this inequality. With the added bound $\tilde{\gamma}$ on $|C x_k - y_{d,k}|$, we see that $y_{c,k}$ necessarily lies between $C x_k - y_{d,k}$ and $\frac{1}{\tilde{\gamma}} (C x_k - y_{d,k})$. Then $y_{c,k} - \frac{1}{\tilde{\gamma}}(C x_k - y_{d,k})$ and $y_{c,k} - (C x_k - y_{d,k})$ must be of opposite signs, hence the inequality.

We thus look for a matrix P such that

$$\sqrt{C P^{-1} C^T} \leq \gamma \tag{7.5}$$

and

$$\left(x_k^T P x_k \leq 1 \wedge y_{d,k}^2 \leq 0.5^2 \wedge (7.4) \right)$$
$$\implies x_{k+1}^T P x_{k+1} \leq 1. \tag{7.6}$$

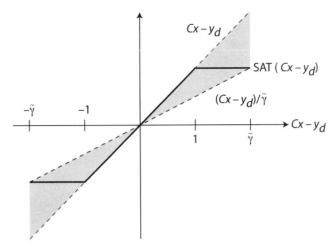

Figure 7.5. Illustration of the sector bound relationship. The equality $y_c =$ SAT$(Cx - y_d)$ (thick line) is abstracted by the inequalities $(Cx - y_d)/\tilde{\gamma} \le y_c \le Cx - y_d$ (gray area).

Defining an extended state vector $\epsilon_k := [x_k\ y_{c,k}\ y_{d,k}\ 1]^{\mathsf{T}}$ and the matrices

$$
\mathcal{U} = \begin{bmatrix} A^{\mathsf{T}}PA & A^{\mathsf{T}}PB & 0_{4\times1} & 0_{4\times1} \\ B^{\mathsf{T}}PA & B^{\mathsf{T}}PB & 0 & 0 \\ 0_{1\times4} & 0 & 0 & 0 \\ 0_{1\times4} & 0 & 0 & -1 \end{bmatrix},
$$

$$
\mathcal{V} := \begin{bmatrix} P & 0_{4\times1} & 0_{4\times1} & 0_{4\times1} \\ 0_{1\times4} & 0 & 0 & 0 \\ 0_{1\times4} & 0 & 0 & 0 \\ 0_{1\times4} & 0 & 0 & -1 \end{bmatrix},
$$

$$
\mathcal{W} := \begin{bmatrix} \frac{2}{\tilde{\gamma}}C^{\mathsf{T}}C & -\left(1+\frac{1}{\tilde{\gamma}}\right)C^{\mathsf{T}} & -\frac{2}{\tilde{\gamma}}C^{\mathsf{T}} & 0_{4\times1} \\ -\left(1+\frac{1}{\tilde{\gamma}}\right)C & 2 & 1+\frac{1}{\tilde{\gamma}} & 0 \\ -\frac{2}{\tilde{\gamma}}C & 1+\frac{1}{\tilde{\gamma}} & \frac{2}{\tilde{\gamma}} & 0 \\ 0_{1\times4} & 0 & 0 & 0 \end{bmatrix},
$$

$$
\mathcal{Y} := \begin{bmatrix} 0_{4\times4} & 0_{4\times1} & 0_{4\times1} & 0_{4\times1} \\ 0_{1\times4} & 0 & 0 & 0 \\ 0_{1\times4} & 0 & 1 & 0 \\ 0_{1\times4} & 0 & 0 & -0.5^2 \end{bmatrix},
$$

we can rewrite Equation (7.6) as

$$
\left(\epsilon_k^{\mathsf{T}}\mathcal{V}\epsilon_k \le 0 \wedge \epsilon_k^{\mathsf{T}}\mathcal{Y}\epsilon_k \le 0 \wedge \epsilon_k^{\mathsf{T}}\mathcal{W}\epsilon_k \le 0\right)
$$
$$
\implies \epsilon_k^{\mathsf{T}}\mathcal{U}\epsilon_k \le 0.
$$

CONVEXIFICATION

Equation (7.6) can then be relaxed by S-Procedure: it will hold if there exists positive coefficients λ, μ, and ν, such that

$$\mathcal{U} - \lambda \mathcal{V} - \mu \mathcal{W} - \nu \mathcal{Y} \preceq 0. \tag{7.7}$$

Equation (7.5) can be rewritten using Schur complement:

$$\begin{bmatrix} \gamma^2 & C \\ C^{\mathsf{T}} & P \end{bmatrix} \preceq 0. \tag{7.8}$$

Note that for fixed λ and γ, Equations (7.7) and (7.8) form a Linear Matrix Inequality (LMI) in P, μ and ν, which means it can be solved by an SDP solver. $\tilde{\gamma} = \gamma + 0.5$ is expected to be larger than 1 (otherwise the saturation would never be activated); moreover, since the saturation should somewhat "bound" this value, we can expect it not to span over multiple orders of magnitude. We also know that $\lambda \in (0, 1)$ thanks to the bottom right coefficient of the LMI (7.7) (since $\nu > 0$). One possible strategy then is to iterate on potential values of λ and γ, solving the corresponding LMI at each iteration. If a solution exists, it will provide the invariant $x^{\mathsf{T}} P x \leq 1$ for the system with saturation.

For our running example, one can generate a suitable template. Values for λ are chosen by exploring $(0, 1)$ with numbers of the form $\frac{k}{2^i}$ for increasing values of $i \geq 1$, and $k < 2^i$. For each choice of λ, the LMI is solved with values of $\tilde{\gamma}$ ranging from 1 to 5 by increments of .1. The solution is found for $\lambda = \frac{63}{64}$ and $\tilde{\gamma} = 3.1$, which amounts to 2605 calls to the LMI solver.

7.1.3.2 First abstracting the disturbance

In the previous approach, the method used was mainly based on an abstraction of the saturation. This second one exposes an alternative method in which the disturbance y_d, rather than the saturation, is abstracted.

Let us first neglect the disturbance y_d and look for a Lyapunov function for the following system:

$$x_{k+1} = \begin{cases} A x_k - B & \text{if } C x_k \leq -0.5 \\ (A + BC) x_k & \text{if } -1.5 \leq C x_k \leq 1.5 \\ A x_k + B & \text{if } C x_k \geq 0.5 \end{cases} \tag{7.9}$$

where A, B, and C are the matrices given in (3.7).

Remark 7.1. y_d *is abstracted in the sense that the term* $(A + BC)x - B y_d$ *of (3.7) is replaced by* $(A + BC)x$ *in (7.9). Similarly, guards such as* $Cx - y_d$ ≤ -1 *are replaced by* $Cx \leq -0.5$ *(since* $|y_d| \leq 0.5$*).*

Remark 7.2. *In case $0.5 \leq \pm Cx_k \leq 1.5$, the system nondeterministically takes one of the two available transitions, the transition taken by the actual system (3.7) being determined by the value of the abstracted variable y_d.*

A quadratic Lyapunov function $x \mapsto x^\mathsf{T} P x$ for this system must then satisfy $x_{k+1}^\mathsf{T} P x_{k+1} \leq x_k^\mathsf{T} P x_k$ for all $x_k \in \mathbb{R}^4$ and all possible transitions from x_k to x_{k+1}. Hence, for all $x \in \mathbb{R}^4$

$$\begin{cases} Cx \leq -0.5 \Rightarrow (Ax - B)^\mathsf{T} P (Ax - B) \leq x^\mathsf{T} P x \\ -1.5 \leq Cx \leq 1.5 \\ \qquad \Rightarrow ((A + BC)x)^\mathsf{T} P ((A + BC)x) \leq x^\mathsf{T} P x \\ Cx \geq 0.5 \Rightarrow (Ax + B)^\mathsf{T} P (Ax + B) \leq x^\mathsf{T} P x. \end{cases}$$

It is worth noting that we can get rid of the first constraint by a symmetry argument. Indeed, the first constraint holds for some x if and only if the third one holds for $-x$. Similarly, we can remove the left part of the implication in the second constraint. Indeed, the right part of the implication holds for some x if and only if it holds for αx and, for α small enough, αx will satisfy the left part of the implication. Thus $x \mapsto x^\mathsf{T} P x$ is a Lyapunov equation for (7.9) if and only if for all $x \in \mathbb{R}^4$

$$\begin{cases} ((A + BC)x)^\mathsf{T} P ((A + BC)x) \leq x^\mathsf{T} P x \\ Cx \geq 0.5 \Rightarrow (Ax + B)^\mathsf{T} P (Ax + B) \leq x^\mathsf{T} P x. \end{cases} \tag{7.10}$$

Defining the vector $x' := [x^\mathsf{T} \ 1]^\mathsf{T}$, this can be rewritten

$$\begin{cases} x^\mathsf{T} (A + BC)^\mathsf{T} P (A + BC) x \leq x^\mathsf{T} P x \\ [C \ 0] \, x' \geq 0.5 \\ \qquad \Rightarrow x'^\mathsf{T} [A \ B]^\mathsf{T} P [A \ B] \, x' \leq x'^\mathsf{T} [I_4 \ 0]^\mathsf{T} P [I_4 \ 0] \, x'. \end{cases}$$

Using a Lagrangian relaxation, this holds when there exists a $\lambda \geq 0$ such that

$$\begin{cases} P - (A + BC)^\mathsf{T} P (A + BC) \succeq 0 \\ [I_4 \ 0]^\mathsf{T} P [I_4 \ 0] - [A \ B]^\mathsf{T} P [A \ B] - \lambda \begin{bmatrix} 0 & C^\mathsf{T} \\ C & -1 \end{bmatrix} \succeq 0 \end{cases}$$

where $M \succeq 0$ means that the matrix M is positive semidefinite (i.e., for all x, $x^\mathsf{T} P x \geq 0$).

We eventually want the template $x^\mathsf{T} P x$ to provide an invariant for the original system with the disturbance y_d. For that purpose, we not only want $(A + BC)^\mathsf{T} P (A + BC)$ in the first inequality to be less than P but rather the least possible, in order to leave some room to later reintroduce y_d. That is, we look for τ_{min}, the least possible $\tau \in (0, 1)$ satisfying

$$\tau P - (A + BC)^{\mathsf{T}} P (A + BC) \succeq 0$$

for some positive definite matrix P. For any given value of τ, this is an LMI and an SDP solver can be used to decide whether a P satisfying it exists or not. Thus, τ_{min} can be efficiently approximated by a bisection search in the interval $(0, 1)$.

Remark 7.3. *τ_{min} is also called minimum decay rate [20].*

We are thus looking for a positive definite matrix P satisfying

$$\begin{cases} \tau_{min} P - (A + BC)^{\mathsf{T}} P (A + BC) \succeq 0 \\ [I_4\,0]^{\mathsf{T}} P [I_4\,0] - [A\,B]^{\mathsf{T}} P [A\,B] - \lambda \begin{bmatrix} 0 & C^{\mathsf{T}} \\ C & -1 \end{bmatrix} \succeq 0. \end{cases}$$

This is an LMI and could then be fed to an SDP solver. Unfortunately, it has no solution. Indeed, A has eigenvalues larger than 1 and taking x large enough can break the second constraint in (7.10) for any value of P.

However, x is saturated when $Cx \geq 1.5$ and it is then reasonable to expect Cx not to go too far beyond this threshold. We thus need to add a constraint $Cx \leq \gamma$ for some $\gamma > 1.5$, in the hope that the generated invariant will eventually satisfy it. This results in the following LMI

$$\begin{cases} \tau_{min} P - (A + BC)^{\mathsf{T}} P (A + BC) \succeq 0 \\ [I_4\,0]^{\mathsf{T}} P [I_4\,0] - [A\,B]^{\mathsf{T}} P [A\,B] - \lambda D \succeq 0 \end{cases} \tag{7.11}$$

where $D := [C \ -0.5]^{\mathsf{T}} [-C\ \gamma] + [-C\ \gamma]^{\mathsf{T}} [C\ -0.5]$.

Finally, for a solution P of the above LMI, $x^{\mathsf{T}} P x \leq r_{max}$ should be a good candidate invariant for the original system (3.7), with $r_{max} := \frac{\gamma^2}{CP^{-1}C^{\mathsf{T}}}$ the largest r such that $x^{\mathsf{T}} P x \leq r$ implies $Cx \leq \gamma$.

On our running example, 15 bisection search iterations first enable us to compute $\tau_{min} = 0.9804$ (rounded to four digits). Then, the values 2, 3, 4, ... are successively tried for γ in (7.11). The LMI appears to have a solution for $\gamma = 2$ and $\gamma = 3$ but not for $\gamma = 4$. The value of P obtained for the last succeeding value of γ ($\gamma = 3$) is then kept as a template and fed to policy iterations along with $r_{max} = 0.26$. All these computations (bisection search for τ_{min}, tests for γ, and computation of r_{max}) took 0.83s on an Intel Core2 @ 1.2GHz.

7.1.3.3 *Relying on computed template*

We use the second method to compute a matrix P:

$$P := \begin{bmatrix} 0.2445 & 0.3298 & -0.0995 & 0.0197 \\ 0.3298 & 1.0000 & -0.0672 & 0.0264 \\ -0.0995 & -0.0672 & 0.0890 & -0.0075 \\ 0.0197 & 0.0264 & -0.0075 & 0.0016 \end{bmatrix}.$$

Then the previous steps can be performed:

Given the templates $t_1 := x^\mathsf{T} P x$, $t_2 := x_{c1}^2$, $t_3 := x_{c2}^2$, $t_4 := x_{p1}^2$, and $t_5 := x_{p2}^2$ where x is the vector $[x_{c1}\ x_{c2}\ x_{p1}\ x_{p2}]^\mathsf{T}$ and (rounded to four digits) policy iterations compute the invariant

$$t_1 \leq 0.1754 \wedge t_2 \leq 6.1265 \wedge t_3 \leq 0.3505$$

$$\wedge\, t_4 \leq 4.1586 \wedge t_5 \leq 1705.1748$$

which implies

$$|x_{c1}| \leq 2.4752 \wedge |x_{c2}| \leq 0.5921 \wedge |x_{p1}| \leq 2.0393 \wedge$$

$$|x_{p2}| \leq 41.2938.$$

Remark 7.4. *Despite the fact that the disturbance y_d was abstracted to generate P, it is worth noting that policy iterations are performed on the complete system, with y_d.*

Remark 7.5. *Although quite heuristic, the choice for γ does not seem that difficult since any value in the interval $(2.40, 3.85)$ would also have led to a good template.*

7.2 ROBUSTNESS WITH VECTOR MARGIN

Beyond stability, an important property which needs to be verified is the robustness of the controller. The property of robustness is necessary in practice as there are many sources of imperfection in the feedback loop. It characterizes "how much" the closed-loop system is stable and which kind of perturbations or uncertainty can be sustained without losing stability. These imperfections can include errors in modeling the plant, uncertainties in the plant that cannot be captured by the model, noises in the sensors and actuators, and limitations of the controller design, i.e., not accounting for the complete range of behaviors of the system, nonlinearities in the actuators, faulty actuators, etc.

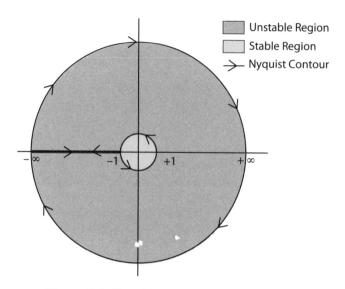

Figure 7.6. Nyquist contour in discrete-time.

The standard metric used in the industry to gauge the robustness of linear SISO[1] controllers consists of phase and gain margins. While these notions are now overseen by more modern techniques such as IQC [25] or μ-analysis [11], they are still widely used in the industry as measures to be guaranteed for a controlled system.

However, these margins are never analyzed or computed on the code artifact, taking into account the real implementation using floating-point arithmetic.

We propose here to rely on the notion of vector margins to characterize such robustness and characterize it as a numerical invariant property over system states, and therefore amenable to code level analysis.

Let us first give an informal overview of classical frequency-based robustness analysis, using Nyquist plots, then present our use of vector margins to bound robustness.

7.2.1 Nyquist plot and stability criterion

The Nyquist plot is the frequency response (magnitude and phase) of the loop transfer function to a sinusoidal input displayed using a polar coordinate system. To construct the Nyquist plot, the loop transfer function $L(z)$ is evaluated along the Nyquist contour Γ. The Nyquist contour, shown in Fig. 7.6, encircles the region outside of the unit disk (OUD) centered at the origin. An example Nyquist plot for the loop transfer function

---

[1]SISO stands for "Single Input Single Output."

$$L(z) := \frac{7.552 \times 10^{-5}z^3 - 7.583 \times 10^{-5}z^2 - 7.454 \times 10^{-5}z + 7.488 \times 10^{-5}}{z^4 - 3.979z^3 + 5.937z^2 - 3.937z + 0.979}$$

$$(7.12)$$

is shown in Fig. 7.7.

We now introduce the Nyquist stability criterion which uses the Nyquist plot to determine the closed-loop stability of the system. Let Z_i be the number of OUD zeros of $L(z) + 1$ and let P_i be the number of OUD poles of $L(z) + 1$. By Cauchy's principle of argument, the Nyquist plot should encircle clockwise[2] the $-1 + 0j$ point N_i number of times where

$$N_i = Z_i - P_i. \qquad (7.13)$$

Using (7.13) and the Nyquist plot in Fig. 7.7, we can conclude the stability of the closed-loop system $\frac{L(z)}{1+L(z)}$ in the following way. First we know the loop transfer function $L(z)$ in (7.12) is stable, i.e., $L(z) + 1$ has 0 OUD poles, which means $P_i = 0$. Since the Nyquist plot in Fig. 7.7 does not encircle $-1 + 0j$, i.e., the critical point, we can conclude that Z_i or the number of OUD zeros of $L(z) + 1$ is also 0. Since OUD zeros of $L(z) + 1$ are also the OUD poles of the closed-loop transfer function, we can conclude that the closed-loop system is also stable.

7.2.2 Phase and gain margins

From the Nyquist stability criterion, one can infer that a possible robustness metric would be the size of the gap between the Nyquist plot and the $-1 + 0j$ point. In fact, phase and gain margins are two different approximations of the "distance" from the Nyquist plot to the critical point.

The first approximation, phase margin, measures how much phase lag the system can tolerate. A phase lag of $\frac{\pi}{2}$ or 90° corresponds to a delay of a quarter of a period. Geometrically speaking, introducing a phase lag of Δ_ω in the feedback loop results in the original Nyquist plot rotated clockwise by Δ_ω, i.e., $L(z) \to e^{j\Delta_\omega} L(z)$. The phase margin represents the amount of clockwise rotation that can be applied to the Nyquist plot before it hits the critical point. As shown in Fig. 7.7, the phase margin (PM) is precisely the clockwise angle between the point where the unit circle, centered at the origin, intersects with the Nyquist plot and $-1 + 0j$.

The second approximation, gain margin, measures how much feedback gain the system can tolerate, i.e., how much one can scale up the Nyquist plot radially before it intersects with the $-1 + 0j$ point. As shown in Fig. 7.7, the gain margin (GM) is precisely $20 \log_{10} \frac{1}{x}$ where x is the magnitude of the Nyquist plot at the phase angle of π. For good robustness, a typical requirement is a phase margin of at least 30° and a gain margin of at least 3db.

---

[2] Counterclockwise encirclement counts as negative.

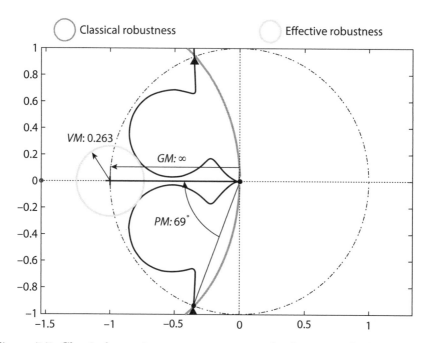

Figure 7.7. Classical margins versus vector margin shown on the Nyquist plot of (7.12).

7.2.3   Vector margin computation

Uncertainties in the feedback loop can introduce simultaneous phase lags and increases in the feedback gain. In those cases, interpreting the phase and gain margins could produce an overly optimistic view of the robustness of the feedback system. For example, a small phase lag combined with a small gain change would destabilize the system in Fig. 7.7, while a pure increase in gain would never do so and it would take a large phase lag alone to destabilize the system. To give a better indication of the robustness of the system, we look at the distance between $-1 + 0j$ and the Nyquist plot induced by the complex modulus, i.e., $\min_{z \in \Gamma} |L(z) + 1|$. In the following, we call this robustness measure the vector margin. By plotting a circle of radius equal to the vector margin centered at the $-1 + 0j$ point, we get the effective robustness envelope in Fig. 7.7, which, for this example, is far more pessimistic than the robustness envelope formed by the classical measures. There are several advantages to using the vector margin.

1. It is a more faithful measure of the robustness.
2. It can be translated into the time-domain and then expressed on the code as a quadratic invariant.
3. It readily extends to MIMO systems [33, 42].

The vector margin can be computed by finding the inverse of the maximum modulus of the sensitivity function $S(z) := \frac{1}{1+L(z)}$ over the Nyquist contour Γ.

This can be seen by noting that

$$\min_{z \in \Gamma} |L(z) + 1| = \cfrac{1}{\displaystyle\max_{z \in \Gamma} \cfrac{1}{|L(z) + 1|}} = \cfrac{1}{\displaystyle\max_{z \in \Gamma} \left| \cfrac{1}{L(z) + 1} \right|}.$$

The sensitivity function is a first-order approximation of the change in the output over the change in the input for the closed-loop system. The state-space representation of the sensitivity function $S(z) := \frac{1}{1+L(z)}$ where $L(z) := P(z)C(z)$ can be expressed in terms of the matrices which form the state-space realization of the plant $P(z)$ and the controller $C(z)$. For the example in Fig. 7.1, the sensitivity transfer function has the following state-space realization

$$
\begin{aligned}
x_{k+1} &= A_s x_k + B_s in_k \\
e_k &= C_s x_k + D_s in_k
\end{aligned}
\tag{7.14}
$$

where

$$
A_s := \begin{bmatrix} A_c & -B_c C_p \\ B_p C_c & A_p - B_p D_c C_p \end{bmatrix} \quad B_s := \begin{bmatrix} B_c \\ B_p D_c \end{bmatrix}
$$

$$
C_s := \begin{bmatrix} 0 & -C_p \end{bmatrix} \quad\quad D_s := \begin{bmatrix} I \end{bmatrix}.
\tag{7.15}
$$

By the application of the bounded real lemma [80, pg. 821], we have the following result.

Property 7.6. *If there exists a positive definite matrix P and $\gamma > 0$, such that*

$$x_{k+1}^\mathsf{T} P x_{k+1} - x_k^\mathsf{T} P x_k \leq \gamma^2 \|in_k\|_2^2 - \|e_k\|_2^2 \tag{7.16}$$

then $\max_{z \in \Gamma} |S(z)| \leq \gamma$.

The inequality in (7.16) is a dissipativity condition [5] and can be checked efficiently by solving a linear matrix inequality (LMI) [23]. We have the following proposition.

Property 7.7. *The previous inequality (7.16) can be written as the following LMI*

$$
\begin{pmatrix} A_s^\mathsf{T} P A_s - P + C_s^\mathsf{T} C_s & A_s^\mathsf{T} P B_s + C_s^\mathsf{T} D_s \\ B_s^\mathsf{T} P A_s + D_s^\mathsf{T} C_s & D_s^\mathsf{T} D_s + B_s^\mathsf{T} P B_s - \gamma^2 I \end{pmatrix} \prec 0.
\tag{7.17}
$$

By Proposition 7.7, for any $P \succ 0$ and $\gamma > 0$ satisfying (7.17), we have $\max_{z \in \Gamma} |S(z)| \leq \gamma$. By minimizing γ in (7.17), we get the vector margin $\delta = \frac{1}{\gamma}$.

Thus, summing (7.16) from time 0 to any time T, we get

$$x_{T+1}^\mathsf{T} P x_{T+1} - x_0^\mathsf{T} P x_0 \leq \gamma^2 \left(\sum_{k=0}^{T} \|in_k\|_2^2 \right) - \sum_{k=0}^{T} \|e_k\|_2^2$$

and since P is positive definite, assuming $x_0 = 0$

$$\sum_{k=0}^{T} \|e_k\|_2^2 \leq \gamma^2 \left(\sum_{k=0}^{T} \|in_k\|_2^2 \right). \tag{7.18}$$

7.2.4 Relationship with phase and gain margins

While vector margins could be computed automatically on the linear system, including its implementation, the use of phase and gain margins is often required to interact with control engineers. We propose here classical projections of vector margins onto a safe approximation of their associated phase and gain margins.

7.2.4.1 Phase margins

As explained in Sec. 7.2.2, the phase margin denotes the angle between the intersection of the Nyquist plot of the transfer function with the unit circle and the point $-1 + 0j$.

This angle is necessarily larger than the angle between the intersection of the computed safe circle of radius δ with the unit circle and the point $-1 + 0j$ (cf. Fig. 7.7, where $\delta = VM$).

In that case a direct projection of vector margins to phase margins is

$$\phi_\delta = 2 \ arcsin(\delta/2).$$

7.2.4.2 Gain margins

Similarly a safe gain margin can be obtained by projecting the vector margin. Gain margin denotes the acceptable scale of the Nyquist plot to avoid intersection with the point $-1 + 0j$.

We can approximate the gain margin associated to the vector margin δ:

$$\Theta_\delta = \frac{1}{1 - \delta}.$$

This gain is usually reported in dB:

$$\Theta_\delta = 20 \cdot log10 \left(\frac{1}{1 - \delta} \right).$$

7.2.5 Spring mass damper analysis

When analyzing the example provided in chapter 3, considering the closed-loop system, one obtains, with classical methods, the following phase and gain margins: $\Theta = 17dB$ and $\phi = 49°$. Note that we assume a system without saturations since saturation cannot be considered when computing Nyquist plot analysis.

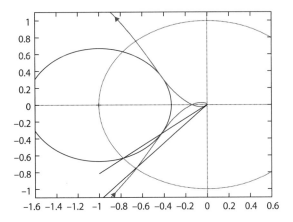

Figure 7.8. Nyquist plot of the spring mass damper system with vector margin.

From the discrete plant and controller description, the sensitivity system is automatically built and analyzed with the LMI (7.17), one obtains $\gamma = 1.4914$, and

$$P = \begin{bmatrix} 111.8330 & 88.4842 & -48.4990 & 8.8432 \\ 88.4842 & 278.5963 & -20.2482 & 6.9605 \\ -48.4990 & -20.2482 & 28.7964 & -3.7961 \\ 8.8432 & 6.9605 & -3.7961 & 0.7013 \end{bmatrix}.$$

The resulting vector margin is $\delta = 1/\gamma = 0.6705$, and its projection to conservative gain and phase margins returns:

$$\Theta_\delta = 10 dB \qquad \phi_\delta = 39°.$$

Fig. 7.8 presents the Nyquist plot and the vector margin.

7.3 RELATED WORK

Few analyses addressed this issue of closed-loop stability and robustness at code level.

At control level, both stability and robustness properties were historically the earliest considered. Lots of techniques address them through different means; we refer the interested (computer scientist) reader to an introductory lecture on control theory [27]. In control theory, two main approaches exist to analyze systems. Either the temporal domain, mentioned above, or the frequency domain, more commonly used. In the frequency domain, stability is usually analyzed by studying the pole placement of the transfer function, either on the Laplace transform of the signal (negative-real part), or on its Z-transform (within the unit circle). In both cases, the system has to be fully linearized (i.e., removing saturation around the linearization point) and the analysis assumes a real semantics, without considering floating-point computations.

Even in the temporal domains analyses, as computed by control theorists, the effect of floating-point computations performed at the controller level and those potentially done during the analysis itself are typically forgotten. These will be addressed in chapter 9.

On the static analysis side, few existing analyses are able to express the simple property of stability. As mentioned earlier, most of the existing abstract domains, used to compute an over-approximation of reachable states, rely on linear approximations. The methods proposed in chapters 5 and 6, providing, respectively, the synthesis of nonlinear sublevel set invariants and narrowing of them using min-policy iterations, addressed this issue. However, as developed in chapter 9, because of floating-point errors a strict inductive condition is enforced, preventing the analysis of saturating controllers. The proposed methods provide additional means to handle this large set of systems.

Finally, a last line of work has to be mentioned: the vast set of work focusing on hybrid systems [75, 95, 105, 125, 127]. It is difficult to summarize in a few words those analyses. We could, however, say that usually (1) they address systems of a somewhat different nature with a central continuous behavior described by differential equations and few discrete events (for instance a bouncing ball or an overflowing water tank) whereas controllers perform discrete transitions on a periodical basis, and (2) they focus on bounded-time properties rather than invariant generation. These two points can be major obstacles to the adaptation of these very interesting techniques to our setting.

For instance, although bounded-time analyses (simulation) do not provide invariants, they enable the use of techniques directly analyzing the continuous plant, such as guaranteed integration [84]. This avoids discretizing the plant, as we do here, which can introduce additional conservatism in the analysis.

Chapter Eight

Validation of System-level Properties at Code Level

ALL PREVIOUS analyses were performed on discrete dynamical systems models. As mentioned in Section 4.1, these models can be either provided as early design artifacts or extracted for more concrete representation such as models or code.

However, once the control-level properties have been expressed and analyzed at model level, we would like to assert their validity on the code artifact extracted from the model.

Luckily, this extraction of code from models is largely automatized thanks to autocoding framework generating embedded code from dataflow models such as MATLAB Simulink, Esterel SCADE, or the academic language LUSTRE [13]. Code generation from dataflow language [77] is now effective and widely used in the industry, supported by tools such as MATLAB Embedded Coder, Esterel KCG, or LUSTRE compilers, e.g., [118].

We claim that code generation can be adapted to enable the expression of system-level properties at code level, and be later proved with respect to the code semantics.

The current chapter addresses this issue. We first give an overview of the modeling framework, enabling the expression of properties at model and code level. A second part explains the generation of such code annotations while a last part focuses on their verification.

8.1 AXIOMATIC SEMANTICS OF CONTROL PROPERTIES THROUGH SYNCHRONOUS OBSERVERS AND HOARE TRIPLES

The framework credible autocoding of control software using control semantics is summarized in Figure 8.1. The framework provides a conduit that allows the domain expert, e.g., the control engineer, to more efficiently produce code with automatically verifiable guarantees of safety and high-level functional properties. Credible autocoding framework adds, on top of the basic model-based development cycle, an additional layer for the generation, translation, and verification of the control semantics of the system. By control semantics of the system, we mean precisely the closed-loop stability and performance properties of the control system and their accompanying proofs. In this framework, the control

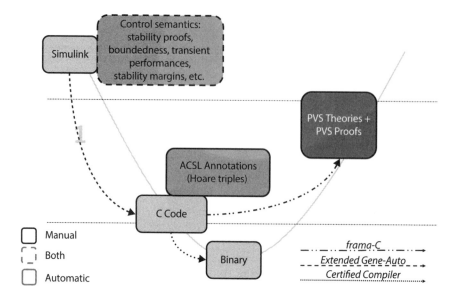

Figure 8.1. Automated credible autocoding and verification framework for control systems.

engineer can choose either to provide manually the properties and proofs as part of the specifications of the controller, or leave it to an automated analyzer that generates the proofs of stability and performance from the controller specifications as presented in chapter 7.

The framework is split into two nominally independent self-contained halves, in which each corresponds to a side in the classic V-diagram of the software development cycle. The left half of the credible autocoding cycle automatically transforms the model of control system into a compilable code annotated with a collection of Hoare logic statements. Taken as a whole, the collection of logic statements, i.e., the annotations along with the code, form a claim of proof that the code satisfies certain closed-loop stability and performance properties of the control system.

The right half of the framework performs an automatic deductive verification of the annotated code with respect to the control system properties expressed by the annotations. The deductive verification process, while assisted by the proof information provided within the generated annotations, is nonetheless independent of the credible autocoding process.

8.1.1   Languages and tools

For the framework, the input language could be any graphical dataflow modeling language such as Simulink or Scicos [59] that is suited for controller design. The input language should have formal and precise semantics so the process for the generation of the code and the control semantics can be formally verified.

The exact choice for the input language is dependent on the domain expert's preference and does not affect the utility of the framework as it can be adapted to other modeling languages such as SCADE. In the rest of the chapter we assume a language such as Simulink for describing models.

Likewise, for the output language, the choice is likely to depend on the preferences of the industry and the certification authority. We chose the language C because of its industrial popularity and the wide availability of static analyzers tailored for C code, including the Frama-C platform [108] from CEA and its weakest precondition analyzer: WP.

The set of annotations in the output source code contains both the functional properties inserted by the domain expert and the proofs that can be used to automatically prove these properties. The analysis of the annotated output could be performed using the formal verification tool Frama-C and the theorem prover PVS. In order to support the automatic proof-checking of the annotated output, a set of linear algebra definitions and theories were integrated into the standard NASA PVS library [114].

In this chapter, the fully automated process from the input model to the verified output is showcased for the property of closed-loop stability, but the expression of other functional properties on the model could also be developed. At this point, we restrict the input to only linear controllers with possible saturations in the loop. For this presentation, examples are related to the spring-mass damper introduced in Section 3.2. However, we have applied the approach to several other much larger systems, which include the Quanser 3-degree-of-freedom helicopter, an industrial F/A-18 UAV controller system and a FADEC control system for a small twin jet turbofan engine [126]. The state-space size of the engine FADEC, for example, is 15.

In the next three sections, we develop our approach on expressing and manipulating control-level semantics at the model in Simulink (see Section 8.1.2), at the code level in C (see Section 8.1.3), and as PVS predicates for the proof part (see Section 8.1.4).

8.1.2 Control semantics in Simulink: ellipsoid-based synchronous observers

When considering synchronous dataflow language a convenient approach to formalize their intended semantics is to rely on synchronous observers [22, 117]. These observers are defined using the modeling language and are included in the description of the system. These are blocks that, depending on internal signals of the system, compute a Boolean output. A valid observation will produce a positive output while a violation of the encoded property will produce a negative one. These blocks characterize a projection, an observation, of the behavior of the system and therefore denote an axiomatic semantics. Synchronous observers can range from simple observation to complex systems in which the block defining the observer is itself defined by numerous subsystems including memories.

These observers can be used for verification and validation. When relying on tests, one can evaluate the output of the block, acting as a test oracle for the encoded property. When performing proof, the goal is to prove that the output of the block is always positive, for all reachable states.

Concerning system-level properties, as we saw in the previous chapter, both boundedness and stability can be expressed using a synchronous observer with inputs $x_i, i = 1, \ldots, n$, and the Boolean-valued function

$$x \to \sum_{i,j=1,\ldots,n} x_i P_{ij} x_j \leq \mu. \tag{8.1}$$

This synchronous observer is parametrized by a symmetric matrix P and a multiplier μ.

For expressing the ellipsoid observers on the Simulink model, we constructed a custom S block denoted as *Ellipsoid* to represent the ellipsoid observer. Additionally, for expressing the operational semantics of the plant, we constructed a custom S block denoted as *Plant*. Its semantics is similar to Simulink's discrete-time state-space block with two key differences. One is that the input to the *Plant* block contains both the input and output of the plant. The other is that the output from the *Plant* block are the internal states of the plant. These variables are used to characterize the current state and are therefore inputs to the ellipsoid observer.

Other properties can also be expressed such as non-expansivity from the dissipativity framework. The *Non-Expansivity* block, when connected with the appropriate inputs and outputs, can be used to express a variety of performance measures such as the H_∞ characteristic of the system or the closed-loop vector margin of the control system. An example of such usage is shown in Figure 8.2 where the closed-loop vector margin of a constant gain controller is expressed using a combination of the *Plant* block and the *Non-Expansivity* block.

In this chapter, we focus on the current fully automated treatment of the open-loop stability properties, hence we will not consider the semantics displayed in Figure 8.2 beyond the description here.

For the running example, we have a Simulink model connected with two synchronous observers. The observers are displayed in the top part of the block diagram for clarity's purpose.

Recall that we made the following assumption for the quantity $y - y_d$:

$$\|y - y_d\| \leq 0.5, \tag{8.2}$$

which is expressed in Figure 8.3 by the Ellipsoid block *BoundedInput* with the parameters $P = 0.5$ and multiplier $\mu = 0.0009$. The stability proof is expressed in Figure 8.3 by the Ellipsoid block *Stability* with the parameters $P = \begin{bmatrix} 6.742 \times 10^{-4} & 4.28 \times 10^{-5} \\ 4.28 \times 10^{-5} & 2.4651 \times 10^{-3} \end{bmatrix}$ and $\mu = 0.9991$. The observer blocks in Figure 8.3 are connected to the model using the *VaMux* block. The role of the *VaMux* block is to concatenate a set of scalar signal inputs into a single vector

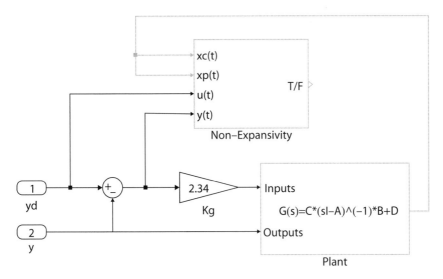

Figure 8.2. Expressing vector margin of the closed-loop system.

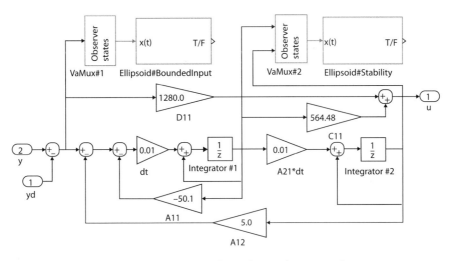

Figure 8.3. Running example with synchronous observers.

output. This special block was constructed because the *Ellipsoid* observer block can accept only a single vector input.

8.1.3 Control semantics at C code level

For the specific problem of open-loop stability, the expressiveness needed at the C code level is twofold. On the one hand, one needs to express that a vector composed of program variables belongs to an ellipsoid: this amounts to a proof of a loop invariant; in the present case, this entails a number of underlying linear algebra concepts. On the other hand, one needs to provide the static analysis tools with indications on how to proceed with the proof of correctness.

The ANSI/ISO C Specification Language (ACSL), is an annotation language for C [148]. It is expressive enough to fulfill our needs, and its associated verification tool, Frama-C [108], offers a wide variety of backend provers which can be used to establish the correctness of the annotated code.

8.1.3.1 Linear algebra in ACSL

A library of ACSL symbols has been developed to express concepts and properties pertaining to linear algebra. In particular, types have been defined for matrices and vectors, and predicates expressing that a vector of variables is a member of the ellipsoid $\mathcal{E}_P$ defined by $\{x \in \mathbb{R}^n : x^\mathsf{T} P x \leq 1\}$, or the ellipsoid $\mathcal{G}_X$ defined by $\left\{ x \in \mathbb{R}^n : \begin{bmatrix} 1 & x^\mathsf{T} \\ x & X \end{bmatrix} \geq 0 \right\}$. For example, expressing that the vector composed of program variables v_1 and v_2 is in the set $\mathcal{E}_P$ where $P = \begin{pmatrix} 1.53 & 10.0 \\ 10.0 & 507 \end{pmatrix}$ can be done with our ACSL extensions using the following annotations:

ACSL
```
/*@ logic matrix P = mat_of_2x2_scalar(1.53,10.0,10
    .0,507);
  @ assert in_ellipsoid(P,vect_of_2_scalar(v_1,v_2)
    ); */
```

The invariance of ellipsoid $\mathcal{E}_P$ throughout any program execution can be expressed by the following *loop invariant*:

C+ACSL
```
//@ loop invariant in_ellipsoid(P,vect_of_2_scalar(
    v_1,v_2));
while (true){
  //loop body
}
```

This annotation expresses that before and after every execution of the loop, the property $\begin{bmatrix} v_1 & v_2 \end{bmatrix}^\mathsf{T} \in \mathcal{E}_P$ will hold. In terms of expressiveness, it is all that is required to express open-loop stability of a linear controller.

However, in order to facilitate the proof, intermediate annotations are added within the loop to propagate the ellipsoid through the different variable assignments, as suggested in [92]. For this reason, a loop body instruction can be annotated with a local contract, like so:

```
──────────────C+ACSL──────────────
/*@ requires in_ellipsoid(P,vect_of_2_scalar(
    v_1,v_2));
  @ ensures in_ellipsoid(Q,vect_of_3_scalar(
      v_1,v_2,v_3));*/
{
// assignment of v_3
}
```

8.1.3.2 Including proof elements

An extension to ACSL, as well as a plug-in to Frama-C, have been developed. They make it possible to indicate the proof steps needed to show the correctness of a contract, by adding extra annotations. For example, the following syntax:

```
──────────────C+ACSL──────────────
/*@ requires in_ellipsoid(P,vect_of_2_scalar(
    v_1,v_2));
  @ ensures in_ellipsoid(Q,vect_of_3_scalar(
      v_1,v_2,v_3));*/
  @ PROOF_TACTIC (use_strategy (AffineEllipsoid));
{
// assignment of v_3
}
```

advises Frama-C to use the strategy `AffineEllipsoid` to prove the correctness of the local contract considered.

8.1.3.3 Closed-loop semantics: expressing plant dynamics in code

In order to express properties pertaining to the closed-loop behavior of the system, one needs to be able to refer to the plant variables. This is achieved through the use of ACSL *ghost* variables. These variables can be initialized and updated like regular C ones, but they only exist in the annotations of the code, and cannot influence the outcome of actual code computations.

This is possible since all our closed-loop analysis, stability, and robustness through vector margins were based on a discrete (linear) representation of the plant semantics. Therefore, the discrete block defining the plant could be used to generate ghost C code.

At the end of the control loop, we use these variables to express the state update of the plant that will result from the computed control signal value. We also enforce axiomatically the fact that the input read from the sensors equals the output of the plant.

8.1.4   Control semantics in PVS

Through a process described in Section 8.3, verifying the correctness of the annotated C code is done with the help of the interactive theorem prover PVS. This type of prover normally relies on a human in the loop to provide the basic steps required to prove a theorem. In order to reason about control systems, linear algebra theories have been developed. General properties of vectors and matrices, as well as theorems specific to this endeavor, have been written and proven manually within the PVS environment.

8.1.4.1   Basic types and theories

Introduced in [114] as part of the larger NASA PVS library, the PVS linear algebra library allows one to reason about matrix and vector quantities, by defining relevant types, operators, and predicates, and proving major properties. To name a few:

- A vector type
- A matrix type, along with all operations relative to the algebra of matrices
- Various matrix subtypes such as square, symmetric, and positive definite matrices
- Block matrices
- Determinants
- High-level results such as the link between Schur's complement and positive definiteness

8.1.4.2   Theorems specific to control theory

In [114], a theorem was introduced, named the ellipsoid theorem. A stronger version of this theorem, along with a couple other useful results in proving open loop stability of a controller, have been added to the library. The first theorem, presented in Fig. 8.4, expresses in the PVS syntax how a generic ellipsoid $\mathcal{G}_Q$ is transformed into $\mathcal{G}_{MQM^\mathsf{T}}$ by the linear mapping $x \mapsto Mx$. A second theorem, presented in Fig. 8.5, expresses how, given 2 vectors x and y in 2 ellipsoids $\mathcal{G}_{Q_1}$ and $\mathcal{G}_{Q_2}$, and multipliers $\lambda_1, \lambda_2 > 0$, such that $\lambda_1 + \lambda_2 \leq 1$, it can always be said that $\begin{pmatrix} x & y \end{pmatrix}^\mathsf{T} \in \mathcal{G}_Q$, where $Q = \begin{pmatrix} \frac{Q_1}{\lambda_1} & 0 \\ 0 & \frac{Q_2}{\lambda_2} \end{pmatrix}$.

```
                            ┤PVS├
ellipsoid_general: THEOREM
 \(\forall\) (n:posnat,m:posnat, Q:SquareMat(n),
        M: Mat(m,n), x:Vector[n], y:Vector[m]):
            in_ellipsoid_Q?(n,Q,x)
            AND y = M*x
        IMPLIES
        in_ellipsoid_Q?(m,M*Q*transpose(M),y)
```

Figure 8.4. Ellipsoid theorem in PVS.

```
                            ┤PVS├
ellipsoid_combination: THEOREM
 \(\forall\) (n,m:posnat, lambda_1, lambda_2:
    posreal, Q_1: Mat(n,n),
        Q_2: Mat(m,m), x:Vector[n], y:Vector[m],
            z:Vector[m+n]):
            in_ellipsoid_Q?(n,Q_1,x)
            AND in_ellipsoid_Q?(m,Q_2,y)
            AND lambda_1+ lambda_2 <= 1
            AND z = Block2V(V2Block(n,m)(x,y))
        IMPLIES
        in_ellipsoid_Q?(n+m,Block2M(M2Block(n,m,n,m
        )(1/lambda_1*Q_1,
                    Zero_mat(m,n),Zero_mat(n,m)
                        ,1/lambda_2*Q_2)),z)
```

Figure 8.5. Ellipsoids combination theorem in PVS.

These two theorems are used heavily in Section 8.3 to prove the correctness of a given Hoare triple. While they are not particularly novel, their proof in PVS was no trivial process and required close to 10000 manual proof steps.

8.2   GENERATING ANNOTATIONS: A STRONGEST POSTCONDITION PROPAGATION ALGORITHM

If provided such powerful tools, one would only need to annotate the loop body with a single loop invariant: the sublevel set μ of the ellipsoid $\mathcal{E}_P$. Unfortunately, because of the complex encoding of the memory model, proof objectives tend to increase with the number of statements considered, and tools have a hard time proving a complex property across a large piece of code.

```
                          ┌─( C+ACSL )─┐
/*@ logic matrix QMat_0 =
    mat_of_2x2_scalar(1484.8760396857954,-25
        .780980284188082,
                -25.780980284188082,406
                    .11067541120576);
*/
/*@ logic matrix QMat_1 =
    mat_of_2x2_scalar(1484.8760396857954,-25
        .780980284188082,
                -25.780980284188082,406
                    .11067541120576);
*/
/*@  logic matrix QMat_2 =
    mat_of_2x2_scalar(1484.8760396857954,-25
        .780980284188082,
                -25.780980284188082,406.11067541120576)
                    ;
*/
```

Figure 8.6. Definition of ellipsoids as annotations.

Therefore, one can introduce intermediate annotations to express locally the impact of the computations to the loop invariant. Ellipsoids characterized by positive definite matrices P cannot contain any dependencies between the variables and the intermediate computation may lead the ellipsoid into a degenerate form. A degenerate ellipsoid can be visualized in two dimensions as a line segment. An alternative way is to manipulate ellipsoids through their Q-form representation, an application of Schur complement.

Definition 8.1 (Q-form). *If matrix P^{-1} exists and let $Q = P^{-1}$, then $x^{\mathsf{T}}Px \leq 1$ is equivalent to $\begin{bmatrix} 1 & x^{\mathsf{T}} \\ x & Q \end{bmatrix} \geq 0$ which is the Q-form.*

When provided with ellipsoids, one can convert it in its Q-form and annotate the loop body–usually a function–with an inductive contract. Then Q-form ellipsoids are injected between each instruction to express the local invariant.

The following figures show a portion of the autocoded output of the running example. The three ACSL annotations in Figure 8.6 define the matrix variables **QMat_0**, **QMat_1**, and **QMat_2**. All three matrix variables parametrize the same ellipsoid as the one obtained from the stability analysis and inserted into the Simulink model as the *Ellipsoid#Stability* observer.

```
/*@
        requires in_ellipsoidQ(QMat_0,
     vect_of_2_scalar(_state_->Integrator_1_memory,
            _state_->Integrator_2_memory));
        requires \valid(_io_) && \valid(_state_);
        ensures in_ellipsoidQ(QMat_1,
     vect_of_2_scalar(_state_->Integrator_1_memory,
     _state_->Integrator_2_memory));
*/
void discrete_timeg_no_plant_08b_compute(
        t_discrete_timeg_no_plant_08b_io *_io_,
        t_discrete_timeg_no_plant_08b_state *
            _state_);
```

Figure 8.7. Loop body function contract.

Using the ACSL keywords *requires* and *ensures*, we can express pre- and post-conditions for lines of code as well as functions. The ACSL function contract in Figure 8.7 expresses the inserted ellipsoid pre- and postconditions for the state-transition function of the controller: **discrete_timeg_no_plant_08b_ compute**.

Within the body of this loop function, ACSL annotations, such as the one in Figure 8.8, manipulate intermediate ellipsoids. In this case, it contains a copy of the precondition from the function contract annotation. This is the inserted ellipsoid precondition for the beginning of the function body.

8.2.1 Invariant forward propagation

The manual forward propagation of ellipsoid invariants was described in [92]. We recall here the basic principles. More details are available in [147] or [145].

8.2.1.1 *Affine transformation*

For the linear propagation of ellipsoids, the *AffineEllipsoid* rule is defined. It is used for linear assignments such as $[\![y]\!] \vDash Lx$, where $x \in \mathbb{R}^m$ is vector of program states and $L \in \mathbb{R}^{1 \times m}$. It is applied to an existing ellipsoid in Q-form, Q_n, and characterizes the next ellipsoid Q_{n+1}.

Recall that Q_n is defined by $\begin{bmatrix} 1 & x^{\mathsf{T}} \\ x & Q \end{bmatrix} \geq 0$ which is equivalent to the more classical inequality $x^{\mathsf{T}} Q^{-1} x \leq 1$.

```
                           ┌─ C+ACSL ─┐
/*@
            behavior ellipsoid0_0:
            requires in_ellipsoidQ(QMat_2,
        vect_of_2_scalar(
            _state_->Integrator_1_memory,
        _state_->Integrator_2_memory));
            ensures in_ellipsoidQ(QMat_3,
        vect_of_3_scalar(
            _state_->Integrator_1_memory,
        _state_->Integrator_2_memory,x1));
            @ PROOF_TACTIC (use_strategy (
                AffineEllipsoid));
      */
{
     x1 = _state_->Integrator_1_memory;
}
```

Figure 8.8. Statement level annotation.

The formal definition of Q_{n+1} involves mechanisms to adapt the dimension of the ellipsoid depending whether y belong or not to the original set of variables in x. However, the main idea is the following: the ellipsoid information obtained after this assignment can be represented by $\begin{bmatrix} 1 & y^{\mathsf{T}} \\ y & LQ_nL^{\mathsf{T}} \end{bmatrix} > 0$. Forward propagation of linear assignment is then easily applicable to Q-form representations.

8.2.1.2   S-Procedure

The *SProcedure* rule is used to combine ellipsoids. We introduced it in Theorem 4.10. In case of multiple ellipsoid invariants to be merged in a single one, we rely on the S-Procedure to compute the multiples and generate the stronger one. This typically arises after a disjunction in the control flow graph such as a saturation.

8.2.2   Verification of the generated postcondition

After the invariant propagation step, we obtain the latest ellipsoid postcondition. It is necessary to check whether this new postcondition implies the inserted ellipsoid precondition defined at the function contract level. Despite numerical error, it is possible to perform a conservative numerical verification by using a safe Cholesky decomposition. This is further developed in chapter 10.

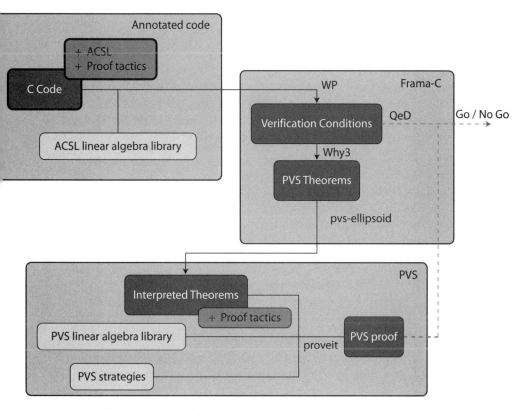

Figure 8.9. General view of the automated verification process. The contribution of this section of the article lies in the domain specific libraries that have been developed at the different layers of description of the code, as well as in the generic proof strategies and the custom Frama-C plug-in pvs-ellipsoid.

8.3 DISCHARGING PROOF OBJECTIVES USING PVS

Once the annotated C code has been generated, it remains to be proven that the annotations are correct with respect to the code. This is achieved by checking that each of the individual Hoare triples hold. Figure 8.9 presents an overview of the checking process. First, the WP plug-in of Frama-C generates verification conditions for each Hoare triple, and discharges the trivial ones with its internal prover QeD. Then, the remaining conditions are translated into PVS theorems for further processing. It is then necessary to match the types and predicates introduced in ACSL to their equivalent representation in PVS. This is done through theory interpretation [41] and outlined in subsection 8.3.2. Once interpreted, the theorems can be generically proven thanks to custom PVS strategies, as described in subsection 8.3.3.

```
                          ┌──────────( C+ACSL )──────────┐
/*@
requires in_ellipsoidQ(QMat_4,
 vect_of_3_scalar(
   _state_->Integrator_1_memory,
   _state_->Integrator_2_memory,
   Integrator_1));
ensures in_ellipsoidQ(QMat_5,
 vect_of_4_scalar(
   _state_->Integrator_1_memory,
   _state_->Integrator_2_memory,
   Integrator_1,
   C11));
PROOF_TACTIC(use_strategy (AffineEllipsoid));
*/
{
    C11 = 564.48 * Integrator_1;
}
```

Figure 8.10. Typical example of an ACSL Hoare triple.

8.3.1   From C code to PVS theorems

The autocoder described in the previous section generates two C functions. One of them is an initialization function, the other implements one execution of the loop that acquires inputs and updates the state variables and the outputs. It is left to the implementer to write the main function combining the two, putting the latter into a loop, and interfacing with sensors and actuators to provide inputs and deliver outputs. Nevertheless, the properties of open- and closed-loop stability, as well as state-boundedness, can be established by solely considering the update function, which this section now focuses on.

Let us consider the annotated code of Figure 8.10. Frama-C/WP is able to characterize the weakest precondition of the ensures statement and to build the proof objective $pre \implies wp(code, post)$. Through the Why3 platform it is able to express it as a PVS theorem. For example, the ACSL/C triple shown in Figure 8.10, taken directly from our running example, becomes the theorem shown in Figure 8.11.

Note that, for the sake of readability, part of the hypotheses of this theorem, including hypotheses on the nature of variables, as well as hypotheses stemming from Hoare triples presented earlier in the code, are omitted here. Note also that in the translation process, functions like malloc_0 or mflt_1 have appeared. They describe the memory state of the program at different execution points.

```
                              ( PVS )
wp: THEOREM
FORALL (integrator_1_0: real):
FORALL (malloc_0: [int -> int]):
FORALL (mflt_2: [addr -> real],
        mflt_1: [addr -> real],
        mflt_0: [addr -> real]):
FORALL (io_2: addr, io_1: addr,
        io_0: addr, state_2: addr,
        state_1: addr, state_0: addr):
...
=> p_in_ellipsoidq(l_qmat_4,
   l_vect_of_3_scalar(
     mflt_2(shift (state_2, 0)),
     mflt_2(shift (state_2, 1)),
     integrator_1_0))
=> p_in_ellipsoidq(l_qmat_5,
   l_vect_of_4_scalar(
     mflt_2(shift (state_2, 0)),
     mflt_2(shift (state_2, 1)),
     integrator_1_0,
     (14112/25 * integrator_1_0)))
```

Figure 8.11. Excerpt of the PVS translation of the triple shown in Figure 8.10.

8.3.2 Theory interpretation

At the ACSL level, a minimal set of linear algebra symbols has been introduced, along with axioms defining their semantics. Section 8.1.4 describes a few of them. Each generated PVS theorem is written within a theory that contains a translation "as is" of these definitions and axioms, along with some constructs specific to handling the semantics of C programs. For example, the ACSL axiom expressing the number of rows of a 2 by 2 matrix:

```
                           ( C+ACSL )
/*@ axiom mat_of_2x2_scalar_row:
  @ {\textbackslash}forall matrix A, real x0101,
      x0102, x0201, x0202;
  @ A == mat_of_2x2_scalar(x0101, x0102, x0201,
      x0202) ==>
  @ mat_row(A) == 2; /*
```

becomes, after translation to PVS:

```
                           ( PVS )
  q_mat_of_2x2_scalar_row :
   AXIOM FORALL (x0101_0:real, x0102_0:real,x0201_0:
      real, x0202_0:real):
            FORALL (a_0:a_matrix):
    (a_0 = l_mat_of_2x2_scalar(x0101_0, x0102_0,
        x0201_0, x0202_0)) =>
    (2 = l_mat_row(a_0))
```

In order to leverage the existing results on matrices and ellipsoids in PVS, theory interpretation is used. It is a logical technique used to relate one axiomatic theory to another. It is used here to map types introduced in ACSL, such as vectors and matrices, to their counterparts in PVS, as well as the operations and predicates on these types. To ensure soundness, PVS requires that what was written as axioms in the ACSL library be proven in the interpreted PVS formalism.

The interpreted symbols and soundness checks are the same for each proof objective, facilitating the mechanization of the process. Syntactically, a new theory is created, in which the theory interpretation is carried out, and the theorem to be proven is automatically rewritten by PVS in terms of its own linear algebra symbols. These manipulations on the generated PVS code are carried out by a Frama-C plug-in called pvs-ellipsoid, which is described below.

8.3.3   Automatizing proofs in PVS

Once the theorem is in its interpreted form, all that remains to do is to apply the proper lemma to the proper arguments. Since we know the theorems used to generate the annotation, as presented in Section 8.2, we provide PVS with strategies of theorems to apply. Without detailing all the issues met when dealing with PVS, one can associate a PVS strategy for each of the proof annotations, both for the affine combination of ellipsoids and the S-Procedure strategy.

8.3.4   Checking inclusion of the propagated ellipsoid

One final verification condition falls under neither AffineEllipsoid nor S- Procedure categories. It expresses that the state remains in the initial ellipsoid $\mathcal{G}_P$. Thanks to a number of transformations, we have proved that the state lies in some ellipsoid $\mathcal{G}'_P$. The conclusion of the verification lies in the final test $P - P' \geq 0$. The current state of the linear algebra library in PVS does not permit us to make such a test, however, a number of very reliable external tools, like the INTLAB package of the MATLAB software suite, can operate this check. One can rely on a sound yet conservative check [116] to ensure positive definiteness of the matrix, with the added benefit of soundness with respect to floating-point computations. These techniques are further developed in chapter 10.

Part IV

Numerical Issues

Chapter Nine

Floating-point Semantics of Analyzed Programs

WHEN IT comes to implementation in a computer, one is limited by the finite bit-level representation of real numbers. A real number is then represented in the computer by a representation in a fixed number of bits. There exist mainly two families of such representations: fixed-point and floating-point representation. Their use largely depends on the application and industrial context and floating-point arithmetics is the main representation used in aerospace applications. This can, however, lead to strange non-expected behaviors (see, e.g., [81] for detailed examples).

Until now all computations and formalizations have assumed real computations. In this chapter a first part outlines floating-point semantics. A second part revisits previous results and adapts them to account for floating-point computations, assuming a bound on the rounding error is provided. A last part focuses on the approaches to bound these imprecisions, over-approximating the floating-point errors.

9.1 FLOATING-POINT SEMANTICS

9.1.1 IEEE 754 floating-point representation

The norm IEEE 754 defines the floating-point representation and operations. In the following we denote by *float* an IEEE 754 floating-point representation. Its implementation can minimally vary but, for the context of this work, we assume that the C level or machine level implementation faithfully respects the norm. On a fixed number of bits–typically 64 bits–one can only represent, as IEEE 754 floating-point values, a finite (but large) number of real numbers. These values are defined as follows:

$$f = (-1)^s \cdot man \cdot 2^{exp} \tag{9.1}$$

where s is bit of sign, man the significand relies on a bit representation over n bits of a positive integer, denoting a fractional part, and exp, the exponent, depends on an unsigned integer representation of k bits. In the following we denote by exp_b the k bits describing the exponent and by man_b the n bits describing the significand. We also denote by $[\![b]\!]^{ui}$ the unsigned integer interpretation of a bit word b.

As an unsigned integer, $[\![exp_b]\!]^{ui}$ ranges in $[0; 2^k - 1]$. Its bounds 0 and $2^k - 1$ denote special values: 0 is used to manipulate floats near zero, we speak about denormalized numbers, while $2^k - 1$ is used for infinities and not-a-number (NaN) values. For all other values of exp_b, the exponent is defined as $[\![exp_b]\!]^{ui} - 2^{k-1} + 1$; we speak about normalized numbers.

Let us first focus on these normalized numbers. In this case, the exponent lives within $[2 - 2^{k-1}, 2^{k-1} - 1]$. In IEEE 754 normalized numbers significand are implicitly prefixed with a leading 1 bit. These $n + 1$ bits characterize the word $[\![1man_b]\!]^{ui} \in [2^n, 2^{n+1} - 1]$. Since the value is interpreted as a significand with only a single leading bit (at the left of the point), it characterizes the set $[2^n, 2^{n+1} - 1] * 2^{-n} = [1, 1 - 2^{-n}]$.

Then, any normalized number can be written as

$$(-1)^s * [\![1man_b]\!]^{ui} * 2^{[\![exp_b]\!]^{ui} - 2^{k-1} - n - 1}.$$

The biggest representable value is then $(2^{n+1} - 1) * 2^{-n} * 2^{(2^k-1) - 2^{k-1} - 1} = (1 - 2^{-n-1}) * 2^{2^{k-1} - 1}$. For 32 bits simple precision float, $k = 8$ and $n = 24$, this gives the maximum value $(1 - 2^{-24}) * 2^{127}$. The smallest positive normal value is obtained with $man_b = 10\ldots0$ and $[\![exp_b]\!]^{ui} = 1$ which evaluates to $(2^n) * 2^{1 - 2^{k-1} + 1 - n} = 2^{2 - 2^{k-1}}$. With simple precision float, this is 2^{-126}.

In case of a tiny value near zero, a rounding error may also produce a false zero value. This is called an underflow. Denormalized numbers are a way to increase the number of floating-point values around zero. They occur when $[\![exp_b]\!]^{ui} = 0$; there is a 0 prefix for significand. In that case the number is written as $(-1)^s * [\![0man_b]\!]^{ui} * 2^{2 - 2^{k-1} - n + 1}$. The smallest positive denormalized value is obtained with $man_b = 0\ldots01$: $2^{2 - 2^{k-1} - n + 1}$. Let eta be such value; with simple precision float, this is $2^{2 - 128 - 23} = 2^{-149}$.

A useful notion is the unit-in-the-last-place (ulp): this function characterizes the distance between two consecutive floats and therefore the maximum imprecision caused by rounding errors. The ulp depends on the exponent part of the representation. For a normalized float value represented as $x = m * 2^e$, with $m = 1, \ldots,$ the minimal distance between two floats is 2^{1-n+e} or 2^{-n+e}. So the ulp of 1 for simple precision is either $2^{-n} = 2^{-23}$ or $2^{1-n} = 2^{-22}$. For double precision, this is 2^{-53} or 2^{-52}. Since it depends on the scale of the value, $\text{ulp}(x) = 2^{e-n} = 2^e * ulp(1) < |x| * ulp(1)$ and can then be directly over-approximated by $|x| * \text{ulp}(1)$. In the following we denote by eps such value ulp(1).

The following table summarizes ulp(1) and minimum positive values for single and double type of floats:

type	bits	k	n	eps $= ulp(1)$	eta
float	32	8	23	$\approx 10^{-7}$	$\approx 10^{-45}$
double	64	11	52	$\approx 10^{-16}$	$\approx 10^{-324}$

9.1.2 Floating-point errors

While computing in floats, rounding errors are introduced and accumulated.

Definition 9.1. *Let $\mathbb{F} \subset \mathbb{R}$ be the set of floating-point values and $fl(e) \in \mathbb{F}$ represents the floating-point evaluation of expression e with any rounding mode and any order of evaluation.*[1]

Example 9.2. *Depending on the order of evaluation, the value $fl(1+2+3)$ can be either $round(1 + round(2+3))$ or $round(round(1+2)+3)$ with round any valid rounding mode (toward $+\infty$ or to nearest for instance), leading to different final values.*

9.1.2.1 Value error

Some simple values such as 0.1 are not exactly representable in floats since they cannot be expressed with a finite number of bits. Each use of a constant may therefore introduce basic errors. The error associated to the constant c is bounded by the minimal distance to the nearest floating-point value, and therefore bounded by $1/2 \operatorname{ulp}(x) < 1/2 |x| \operatorname{ulp}(1)$.

9.1.2.2 Numerical operations: addition and multiplication

Similarly, each numerical operation introduces errors. For example, since the addition of two floats may not be exactly representable in floats, the result is rounded to a floating-point value, depending on the rounding mode.

Let $e^+(u, v)$ be such additive error. In case of u and v associated to their existing error e^u and e^v, the computation of $(u + e^u) + (v + e^v)$ returns $(u + v) + (e^u + e^v + e^+(u, v))$.

More practically it is possible to provide a bound for such error using eps:

Definition 9.3. *Let* eps $= \operatorname{ulp}(1)$ *is the precision of the floating-point format $\mathbb{F}$. We have for all $x, y \in \mathbb{F}$*

$$\exists \delta \in \mathbb{R}, |\delta| \leq \text{eps}$$
$$fl(x+y) = (1 + \delta)(x + y). \tag{9.2}$$

Concerning multiplication, similar errors are introduced. We denote by $e^\times(u, v)$ such multiplicative error. When considering input with existing errors e^u and e^v, the computation of $(u + e^u) * (v + e^v)$ returns $(u * v) + (e^u * v + e^v * u + e^*(u, v))$.

---

[1] *Order of evaluation matters since floating-point addition is not associative.*

Similarly, this error can be bounded using both the floating-point precision eps and the underflow constant eta:

Definition 9.4. *Let* eps *be the precision of the floating-point format* $\mathbb{F}$ *and* eta *the precision in case of underflows. In particular, we have for all* $x, y \in \mathbb{F}$

$$\exists \delta, \eta \in \mathbb{R}, |\delta| \leq \mathsf{eps} \wedge |\eta| \leq \mathsf{eta}$$
$$fl(x \times y) = (1 + \delta)(x \times y) + \eta. \tag{9.3}$$

Remark 9.5. *Those are fairly classic notations and results [26, 96].*

9.2 REVISITING INDUCTIVENESS CONSTRAINTS

Taking floating-point arithmetic into account, one needs to reevaluate the constraints used in the previous chapters to account for floating-point noise.

Let us be given a function that, provided bounds over input variables x, compute a safe over-approximation of the floating-point error $err_f > 0$ when computing, in floating-point arithmetics, $fl(f(x))$.

The semantics of our discrete dynamical systems, as introduced in Section 4.1, were defined using a linear function $f(x, u) = Ax + Bu + b$, in Sections 4.1.1 and 4.1.2, or a polynomial function $T(x)$ for polynomial systems in Section 4.1.3. In addition, piecewise systems, linear or polynomial ones, were constrained by guards: $\bigwedge_i r^i(x) \leq b^i$ with r^i at most quadratic for piecewise linear systems.

9.2.1 Lyapunov conditions with floats

The Lyapunov condition (5.7) imposes that $V \circ f(x) - V(x) \leq 0$. When considering a floating-point implementation, we need to check this other constraint $V \circ (fl(f(x))) - V(x) \leq 0$. It can be reinforced by considering an upper bound $err_{V \circ f}$ on the floating-point errors resulting from $fl(f(x))$ and propagated through function V, such that

$$(V \circ f(x) - V(x) + err_{V \circ f} \leq 0) \Rightarrow (V \circ (fl(f(x)) - V(x) \leq 0).$$

Note that, since $V(x) = p(x)$ or $V(x) = x^\mathsf{T} P x$ are at least quadratic polynomials, the error obtained is not linear in the initial floating-point error of $fl(f(x))$. Let $err_{V \circ f}$ be such error.

FLOATING-POINT ASSIGNMENTS

All inductiveness equations of chapter 5 can be adapted to account for these floating-point errors: linear systems (5.9) and (5.10); preserving the shape (5.11); considering inputs (5.14) while bounding $f(x) = Ax + Bu$, with both x and u bounded; piecewise linear systems with their invariance constraint (5.27)

where the term $-F^{i\mathsf{T}} \begin{pmatrix} 0 & q^j\mathsf{T} \\ q^j & P^j \end{pmatrix} F^i$ shall be replaced by $-F^{i\mathsf{T}} \begin{pmatrix} 0 & q^j\mathsf{T} \\ q^j & P^j \end{pmatrix} F^i -$
$err_{F^{i\mathsf{T}}P^jF^i}$, and similarly for their k-inductive extension: Eqs. (5.46) and (5.47); and lastly, polynomial systems with the constraint $p - p \circ T^i$ in Eq. (5.56) are substituted by $p - p \circ T^i - err_{p \circ T^i}$.[2]

FLOATING-POINT GUARDS

All methods proposed for piecewise systems integrate the condition in the constraints. Both quadratic constraints in the linear and more general polynomial ones are encoded as negativity constraints and introduced in the inductiveness constraint through a Lagrangian or SOS relaxation. Similar to the assignment computed in floating-point arithmetic, the evaluation of the guard $r(x) \leq 0$ becomes $fl(r(x)) \leq 0 \equiv r(x) \pm err_r \leq 0$ with err_r the actual floating-point error. In case of a bound on variable x this quantity err_r can be over-approximated. Let $\overline{err_r}$ be such upper bound. One can therefore encode the constraint by the stronger condition $r(x) - \overline{err_r} \leq 0$. If this condition holds, then the computation with floats of the guard $r(x) \leq 0$ will also hold. Note that, because of this stronger encoding, these sound guards do not characterize anymore a partitioning of the state space: multiple transitions could be computed from the same value. This would correspond to a nondeterministic abstract of the initial deterministic partitioned dynamical system.

In both cases, assignments and guards, a coarse over-approximation of the floating-point error err will only over-constrain the result: more feasible transitions, and stricter inductiveness conditions.

9.2.2 Policy iteration

In the policy iteration algorithm presented in chapter 6, the computation of an individual policy, in Eq. (6.3) and Eq. (6.5) for the SOS version, admits similar issues: $p \circ T^i$ should account for floating-point errors in the computation of T^i while $\langle \mu, r^i \rangle$ should consider floating-point errors in guards. Thanks to the available templates $q \in \mathbb{P}$, one can bound the variable x and over-approximate these two floating errors: $err_{p \circ T^i}$ and err_{r^i}. The constraint can be weakened on the guards side and strengthened on the assignment side:

$$\left(F_i^{\mathcal{R}}(w)\right)(p) = p \circ T^i + err_{p \circ T^i} + \sum_{q \in \mathbb{P}} \lambda(q)(w(q) - q) - \langle \mu, r^i - err_{r^i} \rangle + \sigma$$

(9.4)

$$\text{and } \deg(\sigma) \leq 2m \deg T^i, \deg(\langle \mu, r^i \rangle) \leq 2m \deg T^i.$$

---

[2]Recall that except the definitions of classical Lyapunov functions for pure linear systems, all our other convexifications represent inductiveness as a positive constraint (rather than a negative one).

9.2.3   System-level analyses

Closed-loop system analyses could be similarly analyzed. However, the plant part is not supposed to generate any floating-point error and could be omitted when computing the bound on the error. Analyses performed at code level (see chapter 8) should compute separately the transformation of the ellipsoid invariants in Q-form, assuming a real semantics, and the approximation of the floating-point errors on the loop body. Then, when checking the final inductiveness of the final ellipsoid in P-form with respect to the initial one, the error has to be considered to further constrain the implication check.

Concerning robustness analysis and vector margins, as presented in Section 7.2, what we need is not exactly the inequality (7.16) describing the dissipativity condition but rather[3]

$$\mathrm{fl}(x_{k+1})^\mathsf{T} P\,\mathrm{fl}(x_{k+1}) - x_k^\mathsf{T} Px_k \leq \gamma^2 \|in_k\|_2^2 - \|\mathrm{fl}(e_k)\|_2^2.$$

Again, bounding by the positive upper bound ϵ the floating-point error occurring when computing the linear update x_{k+1} evaluated through the Lyapunov function, one can express the stronger condition:

$$x_{k+1}^\mathsf{T} P\,x_{k+1} - x_k^\mathsf{T} Px_k + \epsilon \leq \gamma^2 \|in_k\|_2^2 - \|e_k\|_2^2.$$

Thus, instead of checking (7.17), we have to check

$$\begin{pmatrix} A_s^\mathsf{T} PA_s - P + C_s^\mathsf{T} C_s + \epsilon I & A_s^\mathsf{T} PB_s + C_s^\mathsf{T} D_s \\ B_s^\mathsf{T} PA_s + D_s^\mathsf{T} C_s & D_s^\mathsf{T} D_s + B_s^\mathsf{T} PB_s - \gamma^2 I + \epsilon I \end{pmatrix} \prec 0.$$

This robustness property is still difficult to prove since the inequality (7.16) is not absolute but relative to the difference between the norm 2 of the input and the error. The inequality should then hold for in_k and e_k as small as possible. However, because of finiteness of floating representation and underflow mechanism, the error they induce is no longer relative but absolute and we can only prove:

$$\mathrm{fl}(x_{k+1})^\mathsf{T} P\,\mathrm{fl}(x_{k+1}) - x_k^\mathsf{T} Px_k$$
$$\leq \gamma^2 \|in_k\|_2^2 - \|\mathrm{fl}(e_k)\|_2^2 + \eta$$

where η is a constant, depending neither on x, nor on in. Thus, instead of (7.18), we get

$$\sum_{k=0}^\mathsf{T} \|e_k\|_2^2 \leq \gamma^2 \left(\sum_{k=0}^\mathsf{T} \|in_k\|_2^2 \right) + (T+1)\eta.$$

---

[3]x_{k+1} and e_k both incur floating-point computations in the controller (cf. (7.14)) whereas in_k is just a real number.

input : Dynamical system defined by function $f(x)$
output: Go/ NoGo
1 Perform a first analysis: synthesize nonlinear template;
2 Rely on policy iteration to compute a bound per local variable;
3 Over-approximate floating-point error on a single loop body evaluation: err_{body};
4 Perform a second analysis: synthesize nonlinear template based on
  $f(x) + err_{body}$;
5 Rely on policy iteration to compute a bound per local variable;
6 Check inclusion in the initial bounds;

Figure 9.1. Algorithm to compute invariant soundness check with respect to floating-point semantics.

However, in practice η is tiny $(\eta \simeq 10^{-324})$ so that it can remain negligible in front of the input in as long as the number T of iterations of the system remains bounded (for instance, the flight commands of a plane typically operate at 100Hz and certainly no longer that 100 hours [54], meaning less than $T := 100 \times 3600 \times 100 \simeq 10^8$ iterations).

9.2.4 Checking soundness of invariant with floating-point error noise

In presence of simple linear updates and guards, assuming conditions on the dimension of the considered systems, and assuming a specific evaluation order, the error on the computation of linear update Ax can be bounded. We refer the reader to the work of Roux [128, 144] for the characterization of such bounds and the formal proof of the results in Coq.

However, in the presence of a more complex dynamic, static analysis provides more flexible means to compute the floating-point error bound. Assuming such analysis is available, we propose the following algorithm to perform this invariant soundness check.

9.3 BOUND FLOATING-POINT ERRORS: TAYLOR-BASED ABSTRACTIONS AKA ZONOTOPIC ABSTRACT DOMAINS

Provided bounds on each variable, computing the floating-point error can be performed with classical interval-based analysis. Kleene based iterations with interval abstract domain provide the appropriate framework to compute such bounds. In our setting this is even simpler since we are interested in bounding the floating-point error on a single call of our dynamic system transition function, that is, a single loop body execution without internal loops.

9.3.1   Interval-based analysis

Classical interval-based abstract domains can be easily adapted to soundly over-approximate floating-point computation by rounding appropriately each operation:

Injection of constant is expressed with the ulp(1) constant:

$$\alpha_{\mathcal{M}}(r) = [fl(r) - |r| * \text{ulp}(1), fl(r) + |r| * \text{ulp}(1)].$$

Note that it is difficult to detect whether a provided value is exactly representable within a given floating representation. The proposed encoding may introduce spurious noise with the $\pm|r| * \text{ulp}(1)$ terms. For example, the exact integer 1 will be encoded by $[r - \text{ulp}(1), r + \text{ulp}(1)]$. A more precise alternative could evaluate the obtained float and the other float obtained with a much larger floating-point representation such as an MPFR float over 1000 bits. In case of similar value, the number is likely (but not proven) to be exactly representable as a float.

Each numerical operation can be adapted to deal with floating-point rounding, rounding the operator to $-\infty$ for the lower bound and to $+\infty$ for the upper bound. We define by $\uparrow_\circ (x)$ the rounding of the real value x towards $\circ \in \{0, +\infty, -\infty, \sim\}$ denoting respectively the rounding towards zero, $+\infty$, $-\infty$, and to the nearest value.

Let us consider a binary operator $\Diamond$ and the resulting interval $[r_1, r_2] = [x_1, x_2] \Diamond [y_1, y_2]$ computed with a real semantics. A sound floating-point interval would then be $[\uparrow_{-\infty} (r_1), \uparrow_{+\infty} (r_2)]$.

The rounding error mode in C can be easily changed but would not be efficient if it has to be changed twice for each operation. One usually chooses the rounding to $+\infty$ and relies on a negation to manipulate lower bounds. We would rather compute $[-\uparrow_{+\infty} (-r_1), \uparrow_{+\infty} (r_2)]$.

For example, for addition we obtain

$$
\begin{aligned}
[x_1, x_2] + [y_1, y_2] = \\
[-\uparrow_{+\infty} (-x_1 - y_1), \uparrow_{+\infty} (x_2 + y_2)].
\end{aligned}
\tag{9.5}
$$

Remember that unary negation only changes the sign bit but does not introduce imprecision. In the following we will rely both on $\uparrow_{+\infty} (x)$ and $\uparrow_{-\infty} (x)$ in the notations, but recall that $\uparrow_{-\infty} (x)$ is implemented as $-\uparrow_{+\infty} (-x)$.

While the previous method computes a sound approximation of floating-point computation, it does not enable the identification of the floating-point error part. An alternative abstraction would rely on rounding to the nearest, and represent a set of floating-point values by a pair of intervals $[f_1, f_2] + [e_1, e_2]$ where the first interval $[f_1, f_2]$ denotes the incorrect bounds obtained when manipulating floats as reals, and the second interval $[e_1, e_2]$ is used to carry on errors. In the following we denote such an element by the 4-tuple (f_1, f_2, e_1, e_2). This new abstract domain is defined over the Cartesian product of two lattices and therefore satisfies all abstract interpretation requirements for an abstract domain: structure lattice, existence of a Galois connection between subset of $\mathbb{R}$ and the domain, etc. We refer the reader to [38, 57, 100] for more details on means to bound these floating-point accumulated rounding errors.

Let us recall the characterization of floating-point values for addition and multiplication presented in Section 9.1.2.2:

$$(u + e^u) + (v + e^v) =$$
$$(u + v) + (e^u + e^v + e^+(u, v)) \tag{9.6}$$

$$(u + e^u) * (v + e^v) =$$
$$(u * v) + (e^u * v + e^v * u + e^*(u, v)) \tag{9.7}$$

with $|e^+(u, v)| \leq |u + v|$ eps and $|e^*(u, v)| \leq |u * v|$ eps + eta.

These equations can be adapted to pairs of intervals as detailed in Figure 9.2.

This method allows us to characterize both the actual values obtained by floating-point computation in the value part and a safe error term. In case of a deterministic loopless code computing an expression exp, one would obtain the abstract value $[x, x] + [e_1, e_2]$ where the singleton interval for the value part denotes exactly the value x that would have been obtained when computing fl(exp). Thanks to the handling of floating-point errors, the computation of exp with reals is guaranteed to lie within $[\uparrow_{-\infty}(x + e_1), \uparrow_{+\infty}(x + e_2)]$.

9.3.2 Affine arithmetic

Affine arithmetic was introduced in the '90s by Comba and Stolfi [21] as an alternative to interval arithmetics, allowing us to avoid some pessimistic computations like the cancellation:

$$x - x = [a, b] -_\mathcal{I} [a, b] = [a - b, b - a] \neq [0, 0].$$

It relies on a representation of convex subsets of $\mathbb{R}$ keeping dependencies between variables: e.g., $x \in [-1, 1]$ will be represented as $0 + 1 * \epsilon_1$ while another variable $y \in [-1, 1]$ will be represented by another ϵ term: $y = 0 + 1 * \epsilon_2$. Therefore $x - x$ will be precisely computed as $\epsilon_1 - \epsilon_1 = 0$ while $x - y$ will result in $\epsilon_1 - \epsilon_2$, i.e., denoting the interval $[-2, 2]$.

Affine arithmetics and variants of it have been studied in the area of applied mathematics community and global optimization. In global optimization, the objective is to precisely compute the minimum or maximum of a nonconvex function, typically using branch and bound algorithms. In most settings the objective function co-domain is $\mathbb{R}$ and interval or affine arithmetics allow us to compute such bounds. Bisection, i.e., branch and bound algorithm, improves the precision by considering subcases. The work of [61] introduced a quadratic extension of affine forms allowing us to express terms in $\epsilon_i \epsilon_j$.

$$(x_1, x_2, e_1, e_2) + (y_1, y_2, f_1, f_2) = \begin{pmatrix} \uparrow_{\sim} (x_1 + y_1), \\ \uparrow_{\sim} (x_2 + y_2), \\ \uparrow_{-\infty} (e_1 + f_1 - e^+(x_1, y_1)), \\ \uparrow_{+\infty} (e_2 + f_2 + e^+(x_2, y_2)) \end{pmatrix} \tag{9.8}$$

$$(x_1, x_2, e_1, e_2) * (y_1, y_2, f_1, f_2) =$$

$$\begin{pmatrix} min(\uparrow_{\sim}(x_1 y_1), \uparrow_{\sim}(x_1 y_2), \uparrow_{\sim}(x_2 y_1), \uparrow_{\sim}(y_1 y_2)), \\ max(\uparrow_{\sim}(x_1 y_1), \uparrow_{\sim}(x_1 y_2), \uparrow_{\sim}(x_2 y_1), \uparrow_{\sim}(y_1 y_2)), \\ \uparrow_{-\infty} \begin{cases} min(\uparrow_{-\infty}(e_1 y_1), \uparrow_{-\infty}(e_1 y_2), \uparrow_{-\infty}(e_2 y_1), \uparrow_{-\infty}(e_2 y_2)) \\ +min(\uparrow_{-\infty}(x_1 f_1), \uparrow_{-\infty}(x_1 f_2), \uparrow_{-\infty}(x_2 f_1), \uparrow_{-\infty}(x_2 f_2)) \\ -min(e^*(x_1, y_1), e^*(x_1, y_2), e^*(x_2, y_1), e^*(x_2, y_2)) \end{cases}, \\ \uparrow_{+\infty} \begin{cases} max(\uparrow_{+\infty}(e_1 y_1), \uparrow_{+\infty}(e_1 y_2), \uparrow_{+\infty}(e_2 y_1), \uparrow_{+\infty}(e_2 y_2)) \\ +min(\uparrow_{+\infty}(x_1 f_1), \uparrow_{+\infty}(x_1 f_2), \uparrow_{+\infty}(x_2 f_1), \uparrow_{+\infty}(x_2 f_2)) \\ +min(e^*(x_1, y_1), e^*(x_1, y_2), e^*(x_2, y_1), e^*(x_2, y_2)) \end{cases} \end{pmatrix}$$
$$\tag{9.9}$$

where $e^+(a, b)$ is defined as $(|a| + |b|)$eps and $e^*(a, b)$ as $|a * b|$ eps + eta.

Figure 9.2. Addition and multiplication on intervals with floating-point errors.

In static analysis, affine forms lifted to abstract environments, as vectors of affine forms, are an interesting alternative to costly relational domains. They provide cheap and scalable relational abstractions: their complexity is linear in the number of error terms–the ϵ_i–while most relational abstract domains have a complexity at least cubic. Since their geometric concretization characterizes a zonotope, i.e., a symmetric convex polytope, they are commonly known as zonotopic abstract domains.

However, since zonotopes are not equipped with a lattice structure, their use in pure abstract interpretation using a Kleene iteration schema is not common. The definition of an abstract domain based on affine forms requires the definition of an upper bound and lower bound operators since no least upper bound and greatest lower bound exist in general. Choices vary from the computation of a precise minimal upper bound to a coarser upper bound that tries to maintain relationship among variables and error terms. For example, the choices of [85] try to compute such bounds while preserving the error terms of the operands, as much as possible, providing a precise way to approximate a functional. In practice convergence of the fixpoint computation is not as easy to guarantee as

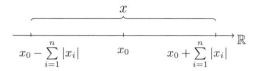

Figure 9.3. Range of an affine form.

it is for (join-)complete lattice. This is, however, not an issue in our context since we do not allow loops within the discrete dynamical system transfer function.

9.3.2.1 *Affine arithmetics: affine forms*

An affine form is characterized by a pair $(c, (b)_m) \in \mathbb{R} \times \mathbb{R}^m$. It defines a map $a \in [-1, 1]^m \to \mathbb{R}$ such that

$$a(\epsilon) = c + b^\mathsf{T}\epsilon.$$

In the following we denote by $\mathbf{A} = \mathbb{R} \times \mathbb{R}^n$ the set of affine forms. The variables $\epsilon_i \in [-1; 1], i \in [1; n]$ are called the error terms.

9.3.2.2 *Geometric interpretation*

An affine form $(c, (b)_n)$ denotes a subset of $\mathbb{R}$. We introduce a concretization function denoting the geometric interpretation of an affine form.

Definition 9.6. *Let* $\gamma_\mathbf{A} : \mathbf{A} \to \wp(\mathbb{R})$ *be the concretization function of* $\mathbf{A}$ *defined as:*

$$\gamma_\mathbf{A} : \qquad \mathbf{A} \quad \to \quad \wp(\mathbb{R})$$

$$(c, (b)_n) \quad \mapsto \quad \left\{ x \in \mathbb{R} \;\middle|\; \begin{array}{c} \exists \epsilon \in [-1; 1]^n, \\ x = c + b^\mathsf{T}\epsilon \end{array} \right\}.$$

When considering an affine form $a \in \mathbf{A}$, one can obtain the set of values $\gamma_\mathbf{A} \in \wp(\mathbb{R})$ it represents by considering only the absolute values of the coefficients, as illustrated in Fig. 9.3:

$$\gamma_\mathbf{A}(c, (b)_m) = \left[c - \Sigma_{i \in [1;m]} |b_i| \,;\, c + \Sigma_{i \in [1;m]} |b_i| \right].$$

9.3.2.3 *Arithmetics*

The set $\mathbf{A}$ is fitted with arithmetic operators: addition, negation, multiplication, and scalar multiplication.

Definition 9.7.

$$
\begin{aligned}
(c, (b)_n) +_{\mathbf{A}} (c', (b')_n) &= (c + c', (b + b')_n) \\
-_{\mathbf{A}}(c, (b)_n) &= (-c, (-b)_n) \\
(c, (b)_n) \times_{\mathbf{A}} (c', (b')_n) &= (c \times c', (b'')_{n+1}) \\
\textit{where} \left\{
\begin{array}{l}
(b'')_n = c \times (b')_n + c' \times (b'')_n \\
b''_{n+1} = \sum_{i=1}^{n} \sum_{i=j}^{n} |b_i \times b'_j|
\end{array}
\right. \\
\lambda *_{\mathbf{A}} (c, (b)_n) &= (\lambda \times c, \lambda \times (b)_n)
\end{aligned}
$$

Note that nonlinear operations introduce a new occurrence of an error term while others are exact.

One can inject an interval into an affine form by introducing a fresh error term ϵ_f:

Definition 9.8. *Let $x = [x^-, x^+]$ be an interval of $\mathbb{R}$. Let $\epsilon_f \in [-1; 1]$ be a fresh error term. We define the affine form associated to x as:*

$$
x = \frac{x^- + x^+}{2} + \frac{x^- - x^+}{2} \epsilon_f.
$$

It characterizes the abstraction function $\alpha_{\mathbf{A}} : \mathbb{R}^2 \to \mathbf{A}$:

$$
\alpha_{\mathbf{A}}(x) = \left(\frac{x^- + x^+}{2}, (\frac{x^- - x^+}{2})_1 \right).
$$

9.3.2.4   Poset structure

While hardly used in the global optimization community where affine forms are used, let us consider a partial order over affine sets. We rely on the geometrical interpretation given by the concretization function $\gamma_{\mathbf{A}}$, and fit the set $\mathbf{A}$ with a poset structure. We define the partial order $\sqsubseteq_{\mathbf{A}}$ as follows:

Definition 9.9. *We define the partial order $\sqsubseteq_{\mathbf{A}} : \mathbf{A} \times \mathbf{A} \to \mathbb{B}$ such that:*

$$
\forall x, y \in \mathbf{A}, x \sqsubseteq_{\mathbf{A}} y \triangleq \gamma_{\mathbf{A}}(x) \subseteq \gamma_{\mathbf{A}}(y).
$$

We also introduce a safe meet operator. Since $\langle \mathbf{A}, \sqsubseteq_{\mathbf{A}} \rangle$ is not fitted with a meet-complete structure, we cannot provide an exact meet operator computing the greatest lower bounds of two affine forms. However, we can rely on an abstract meet which provides a safe but imprecise upper bound of maximal lower bounds.

The following function performs such computation: it projects each affine form to its interval representation on which it performs the meet computation, before abstracting again to a fresh affine form.

Definition 9.10. *Let $(c, (b)_n)$ and $(c', (b')_n) \in \mathbf{A}$. We define the meet operator $\sqcap_{\mathbf{A}} : \mathbf{A} \times \mathbf{A} \to \mathbf{A}$ such that*

$$(c, (b)_n) \sqcap_{\mathbf{A}} (c', (b')_n) =$$
$$\begin{cases} (c, (b)_n) & when\ (c, (b)_n) = (c', (b')_n), \\ \alpha_{\mathbf{A}}(\gamma_{\mathbf{A}}(c, (b)_n) \cap \gamma_{\mathbf{A}}(c', (b')_n)) & otherwise. \end{cases}$$

9.3.2.5   Floating-point error computation

Floating-point errors can be carried in identified error terms. Floating-point error terms can be merged in order to make sure that the number of associated error terms does not increase significantly. However, some specific floating-point errors can be identified such as second-order combinations of floating-point error terms.

9.3.3   Quadratic extension of zonotopes

9.3.3.1   Quadratic forms

An interesting extension of affine arithmetics is quadratic arithmetics [61]. It is a comparable representation of values fitted with similar arithmetics operators but quadratic forms also consider products of two error terms $\epsilon_i \epsilon_j$. A quadratic form is also parameterized by additional error terms used to encode nonlinear errors: $\epsilon_\pm \in [-1, 1], \epsilon_+ \in [0, 1]$ and $\epsilon_- \in [-1, 0]$. Let us define the set $\mathbf{C}^m \triangleq [-1, 1]^m \times [-1, 1] \times [0, 1] \times [-1, 0]$. A quadratic form on m noise symbols is a function q from $\mathbf{C}^m$ to $\mathbb{R}$ defined for all $t = (\epsilon, \epsilon_\pm, \epsilon_+, \epsilon_-) \in \mathbf{C}^m$ by $q(t) = c + b^{\mathsf{T}} \epsilon + \epsilon^{\mathsf{T}} A \epsilon + c_\pm \epsilon_\pm + c_- \epsilon_- + c_+ \epsilon_+$. The A term will generate the quadratic expressions in $\epsilon_i \epsilon_j$. A quadratic form is thus characterized by a 6-tuple $(c, (b)_m, (A)_{m^2}, c_\pm, c_+, c_-) \in \mathbb{R} \times \mathbb{R}^m \times \mathbb{R}^{m \times m} \times \mathbb{R}_+ \times \mathbb{R}_+ \times \mathbb{R}_+$. Without loss of generality, the matrix A can be assumed symmetric. To simplify, we will use the terminology quadratic form for both the function defined on $\mathbf{C}^m$ and the 6-tuple. Let $\mathcal{Q}^m$ denote the set of quadratic forms.

9.3.3.2   Geometric interpretation

Let $q \in \mathcal{Q}^m$. Since q is continuous, the image of $\mathbf{C}^m$ by q is a closed bounded interval. In our context, the image of $\mathbf{C}^m$ by q defines its geometric interpretation.

Definition 9.11. *The concretization map of a quadratic form $\gamma_{\mathcal{Q}} : \mathcal{Q}^m \to \wp(\mathbb{R})$ is defined by:*

$$\gamma_{\mathcal{Q}}(q) = \{x \in \mathbb{R} \mid \exists t \in \mathbf{C}^m\ \text{s.t.}\ x = q(t)\}.$$

Remark 9.12. *We can have $\gamma_{\mathcal{Q}}(q) = \gamma_{\mathcal{Q}}(q')$ with $q \neq q'$, e.g., $q = \epsilon_1^2$ and $q' = \epsilon_2^2$. Therefore, $\gamma_{\mathcal{Q}}$ could not characterize an antisymmetric relation on $\mathcal{Q}^m$ and* *therefore not a partial order. We could consider equivalence classes instead to get an order but we would lose the information that q_1 and q_2 are not correlated.*

The projection of q to intervals consists in computing the infimum and the supremum of q over $\mathbf{C}^m$, i.e., the values:

$$\mathbf{b}^q \triangleq \inf\{q(x) \mid x \in \mathbf{C}^m\} \tag{9.10}$$

$$\mathbf{B}^q \triangleq \sup\{q(x) \mid x \in \mathbf{C}^m\}. \tag{9.11}$$

Computing $\mathbf{b}^q$ and $\mathbf{B}^q$ is reduced to solving a nonconvex quadratic problem which is NP-hard [17]. The approach described in [61] uses simple inequalities to give a safe over-approximation of $\gamma_{\mathcal{Q}}(q)$. The interval provided by this approach is $[\mathbf{b}^q_{MT}, \mathbf{B}^q_{MT}]$ which is defined as follows:

$$
\begin{cases}
\mathbf{b}^q_{MT} \triangleq \begin{aligned}[t] c - \sum_{i=1}^{m} |b_i| - \sum_{\substack{i,j=1,\dots,m \\ j \neq i}} |A_{ij}| + \sum_{i=1}^{m} [A_{ii}]^{-} \\ -c_{-} - c_{\pm} \end{aligned} \\[2em]
\mathbf{B}^q_{MT} \triangleq \begin{aligned}[t] c + \sum_{i=1}^{m} |b_i| + \sum_{\substack{i,j=1,\dots,m \\ j \neq i}} |A_{ij}| + \sum_{i=1}^{m} [A_{ii}]^{+} \\ +c_{+} + c_{\pm} \end{aligned}
\end{cases}
\tag{9.12}
$$

where for all $x \in \mathbb{R}$, $[x]^{+} = x$ if $x > 0$ and 0 otherwise and $[x]^{-} = x$ if $x < 0$ and 0 otherwise.

In practice, we use the interval projection operator $\mathcal{P}^{MT}_{\mathcal{Q}}(q) \triangleq [\mathbf{b}^q_{MT}, \mathbf{B}^q_{MT}]$ instead of $\gamma_{\mathcal{Q}}(q)$, since $\gamma_{\mathcal{Q}}(q) \subseteq \gamma_I\left(\mathcal{P}^{MT}_{\mathcal{Q}}(q)\right)$ where γ_I denotes the concretization of intervals. In [138], we presented a tighter over-approximation of $\gamma_{\mathcal{Q}}(q)$ using SDP.

We will need a "reverse" map to the concretization map $\gamma_{\mathcal{Q}}$: a map which associates a quadratic form to an interval. We call this map the *abstraction map*. Note that the abstraction map produces a fresh noise symbol.

First, we introduce some notations for intervals. Let $\mathcal{I}$ be the set of closed bounded real intervals, i.e., $\{[a, b] \mid a, b \in \mathbb{R}, a \leq b\}$ and $\overline{\mathcal{I}}$ its unbounded extension, i.e., $a \in \mathbb{R} \cup \{-\infty\}, b \in \mathbb{R} \cup \{+\infty\}$. $\forall [a, b] \in \mathcal{I}$, we define two functions $\lg([a, b]) = (b - a)/2$ and $\mathrm{mid}([a, b]) = (b + a)/2$. Let $\sqcup_{\mathcal{I}}$ be the classic join of $\mathcal{I}$ that is $[a, b] \sqcup_{\mathcal{I}} [c, d] \triangleq [\min(a, c), \max(b, d)]$. Let $\sqcap_{\overline{\mathcal{I}}}$ be the classic meet of intervals.

Definition 9.13. *The abstraction map* $\alpha_{\mathcal{Q}} : \mathcal{I} \to \mathcal{Q}^1$ *is defined by:*

$$\alpha_{\mathcal{Q}}([a_1, a_2]) = (c, (b)_1, (0)_1, 0, 0, 0)$$

where $c = \mathrm{mid}\left([a_1, a_2]\right)$ *and* $b = \lg\left([a_1, a_2]\right)$.

Property 9.14 (Concretization of abstraction).

$$\gamma_{\mathcal{Q}}\left(\alpha_{\mathcal{Q}}\left([a_1, a_2]\right)\right) \supseteq [a_1, a_2]$$

9.3.3.3   Arithmetic operators

Quadratic forms are equipped with arithmetic operators whose complexity is quadratic in the number of error terms. We give here the definitions of the arithmetics operators:

Definition 9.15.   *Addition, negation, and multiplication by scalar are defined by:*

$$
\begin{aligned}
&(c, (b)_m, (A)_{m^2}, c_\pm, c_+, c_-) \\
+_{\mathcal{Q}}&(c', (b')_m, (A')_{m^2}, c'_\pm, c'_+, c'_-) = \\
&\left(
\begin{array}{l}
c + c', (b + b')_m, (A + A')_{m^2}, \\
c_\pm + c'_\pm, c_+ + c'_+, c_- + c'_-
\end{array}
\right)
\end{aligned}
\tag{9.13}
$$

$$
\begin{aligned}
-_{\mathcal{Q}}&(c, (b)_m, (A)_{m^2}, c_\pm, c_+, c_-) = \\
&(-c, (-b)_m, (-A)_{m^2}, c_\pm, c_-, c_+)
\end{aligned}
\tag{9.14}
$$

$$
\begin{aligned}
\lambda *_{\mathcal{Q}}&\ (c, (b)_m, (A)_{m^2}, c_\pm, c_+, c_-) = \\
&(\lambda c, \lambda(b)_m, \lambda(A)_{m^2}, |\lambda| c_\pm, |\lambda| c_+, |\lambda| c_-).
\end{aligned}
\tag{9.15}
$$

The multiplication is more complex since it introduces additional errors:

$$
\begin{aligned}
&(c, (b)_m, (A)_{m^2}, c_\pm, c_+, c_-) \\
\times_{\mathcal{Q}}&(c', (b')_m, (A')_{m^2}, c'_\pm, c'_+, c'_-) = \\
\left\{
\begin{array}{l}
\left(
\begin{array}{l}
cc', c'(b)_m + c(b')_m, \\
c'(A)_{m^2} + c(A')_{m^2} + (b)_m (b')_m^{\mathsf{T}}, \\
c''_\pm, c''_+, c''_-
\end{array}
\right) \quad with \\
c''_x = c''_{x_1} + c''_{x_2} + c''_{x_3} + c''_{x_4}, \forall x \in \{+, -, \pm\}.
\end{array}
\right.
\end{aligned}
\tag{9.16}
$$

Each c''_{x_i} accounts for multiplicative errors with more than quadratic degree, obtained in the following four subterms: (1) $\epsilon^{\mathsf{T}} A \epsilon \times \epsilon^{\mathsf{T}} A' \epsilon$; (2) $b^{\mathsf{T}} \epsilon \times \epsilon^{\mathsf{T}} A' \epsilon$ and $b'^{\mathsf{T}} \epsilon \times \epsilon^{\mathsf{T}} A \epsilon$; (3) multiplication of a matrix element in A, A' times an error term in $\pm, +, -$; and (4) multiplication between error terms or with constant c, c'. Their precise definition can be found in [61, §3].

9.3.3.4   Quadratic zonotopes: a zonotopic extension of quadratic forms to environments

Quadratic vectors are the lift to environments of quadratic forms. They provide a p-dimensional environment in which each dimension/variable is associated to

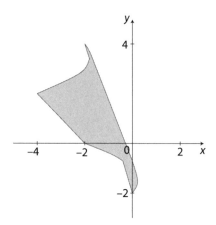

Figure 9.4. Zonotopic concretization of the quadratic vector $q \in \mathcal{Z}^p_{Q^m}$ of Ex. 9.16: $\gamma_{\mathcal{Z}_Q}(q)$.

a quadratic form. As for the affine sets used in zonotopic domains [86], the different variables share (some) error terms; this characterizes a set of relationships between variables, when varying the values of ϵ within $[-1, 1]^m$. The geometric interpretation of quadratic vectors are nonconvex, nonsymmetric subsets of $\mathbb{R}^p$. In the following, we call them quadratic zonotopes to preserve the analogy with affine sets and zonotopes.

Example 9.16 (Quadratic vector).   *Let us consider the following quadratic vector q:*

$$q = \begin{cases} x = -1 + \epsilon_1 - \epsilon_2 - \epsilon_{1,1} \\ y = 1 + 2\epsilon_2 + \epsilon_{1,2}. \end{cases}$$

Note that it corresponds to the following vector of tuples defined over the sequence (ϵ_1, ϵ_2) of error terms:

$$\begin{cases} x = \left(-1, (1, -1)^\mathsf{T}, \begin{pmatrix} -1 & 0 \\ 0 & 0 \end{pmatrix}, 0, 0, 0\right) \\ y = \left(1, (0, 2)^\mathsf{T}, \begin{pmatrix} 0 & 1/2 \\ 1/2 & 0 \end{pmatrix}, 0, 0, 0\right). \end{cases}$$

Fig. 9.4 represents its associated geometric interpretation, a quadratic zonotope.

Let $\mathcal{Z}^p_{Q^m}$ be the set of quadratic vectors of dimension p: $(q^p) \in \mathcal{Z}^p_{Q^m} = (c^p, (b)^p_m, (A)^p_{m^2}, c^p_\pm, c^p_+, c^p_-) \in \mathbb{R}^p \times \mathbb{R}^{p \times m} \times \mathbb{R}^{p \times m \times m} \times \mathbb{R}^p_+ \times \mathbb{R}^p_+ \times \mathbb{R}^p_+$.

The zonotope domain is then a parametric relational abstract domain, parametrized by the vector of m error terms. In practice, its definition mimics a nonrelational domain based on an abstraction $\mathcal{Z}^p_{Q^m}$ of $\wp(\mathbb{R}^p)$. Operators are (i) assignment of a variable of the zonotope to a new value defined by an arithmetic expression, using the semantics evaluation of expressions in $\mathcal{Q}$ and the substitution in the quadratic vector; and (ii) guard evaluation, i.e., constraint over a zonotope, using the classical combination of forward and backward evaluations of expressions [49, §2.4.4].

GEOMETRIC INTERPRETATION AND BOX PROJECTION

One can consider the geometric interpretation as the concretization of a quadratic vector to a quadratic zonotope.

From now on, for all $n \in \mathbb{N}$, $[n]$ denotes the set of integers $\{1, \ldots, n\}$.

Definition 9.17.   *The concretization map* $\gamma_{\mathcal{Z}_\mathcal{Q}} : \mathcal{Z}^p_{\mathcal{Q}^m} \mapsto \wp(\mathbb{R}^p)$ *is defined for all* $q = (q_1, \ldots, q_p) \in \mathcal{Z}^p_{\mathcal{Q}^m}$ *by:*

$$\gamma_{\mathcal{Z}_\mathcal{Q}}(q) = \left\{ x \in \mathbb{R}^p \;\middle|\; \begin{array}{l} \exists\, t \in \mathbf{C}^m \text{ s.t.} \\ \forall\, k \in [p], \; x_k = q_k(t) \end{array} \right\}.$$

Remark 9.18.   *Characterizing such subset of* $\mathbb{R}^p$ *explicitly as a set of constraints is not easy. A classical (affine) zonotope is the image of a polyhedron (hypercube) by an affine map, hence it is a polyhedron and can be represented by a conjunction of affine inequalities. For quadratic vectors a representation in terms of conjunction of quadratic or at most polynomial inequalities is not proven to exist. This makes the concretization of a quadratic set difficult to compute precisely.*

To ease the latter interpretation of computed values, we rely on a naive projection to boxes: each quadratic form of the quadratic vector is concretized as an interval using $\gamma_\mathcal{Q}$.

PREORDER STRUCTURE AND UPPER BOUND

We equip quadratic vectors with a preorder relying on the geometric inclusion provided by the map $\gamma_{\mathcal{Z}_\mathcal{Q}}$.

Definition 9.19.   *The preorder* $\sqsubseteq_{\mathcal{Z}_\mathcal{Q}}$ *over* $\mathcal{Z}^p_{\mathcal{Q}^m}$ *is defined by:*

$$x \sqsubseteq_{\mathcal{Z}_\mathcal{Q}} y \iff \gamma_{\mathcal{Z}_\mathcal{Q}}(x) \subseteq \gamma_{\mathcal{Z}_\mathcal{Q}}(y) .$$

Remark 9.20.   *Since* $\gamma_{\mathcal{Z}_\mathcal{Q}}$ *is not computable,* $x \sqsubseteq_{\mathcal{Z}_\mathcal{Q}} y$ *is not decidable. Note also that, from Remark 9.12, the binary relation* $\sqsubseteq_{\mathcal{Z}_\mathcal{Q}}$ *cannot be antisymmetric and thus cannot be an order.*

Remark 9.21. *The least upper bound of* $Z \subseteq \mathcal{Z}^p_{\mathcal{Q}^m}$, *i.e., an element* z' *s.t.*

$$\forall z \in Z, z \sqsubseteq_{\mathcal{Z}_{\mathcal{Q}}} z' \land$$
$$\begin{pmatrix} \forall z \in Z, \forall z'' \in \mathcal{Z}^p_{\mathcal{Q}^m} \\ z \sqsubseteq_{\mathcal{Z}_{\mathcal{Q}}} z'' \end{pmatrix} \implies z' \sqsubseteq_{\mathcal{Z}_{\mathcal{Q}}} z''$$

does not necessarily exist.

Related work [85, 86, 94, 112, 113] addressed this issue by providing various flavors of join operator computing a safe upper bound or a minimal upper bound. Classical Kleene iteration scheme was adapted[4] to fit this loose framework without (efficient) least upper bound computation. Note that, in general, the aforementioned zonotopic domains do not rely on the geometric interpretation as the concretization to $\wp(\mathbb{R})$.

We now detail a join operator. It is the lifting of the operator proposed in [86] to quadratic vectors. The motivation of this operator is to provide an upper bound while minimizing the set of error terms lost in the computation.

First, we introduce a useful function *argmin*: it cancels values of opposite sign but provides the argument with the minimal absolute value when provided with two values of the same sign:

Definition 9.22. *We define for all* $a \in \mathbb{R}$, $\operatorname{sgn}(a) = 1$ *if* $a \geq 0$ *and -1 otherwise. The* argmin *function*, $\operatorname{argmin} : \mathbb{R} \times \mathbb{R} \to \mathbb{R}$, *is defined as:* $\forall a, b \in \mathbb{R}, \operatorname{argmin}(a, b) = \operatorname{sgn}(a) \min(|a|, |b|)$ *if* $ab \geq 0$ *and 0 otherwise.*

We also need the projection map which selects a specific coordinate of a quadratic vector.

Definition 9.23. $\forall k \in [p]$, *the family of projection maps* $\pi_k : \mathcal{Z}^p_{\mathcal{Q}^m} \to \mathcal{Q}^m$, *is defined by:* $\forall q = (q_1, \ldots, q_p) \in \mathcal{Z}^p_{\mathcal{Q}^m}$, $\pi_k(q) = q_k$.

When a quadratic form q is defined before a new noise symbol is created, we have to extend q to take into account this fresh noise symbol. We introduce an extension map operator that extends the size of the error term vector considered. Informally, $\operatorname{ext}_{i,j}(q)$ adds i null error terms at the beginning of the error term vector and j at its tail, while keeping the existing symbols in the middle.

Definition 9.24. *Let* $i, j \in \mathbb{N}$. *The extension map* $\operatorname{ext}_{i,j} : \mathcal{Q}^m \to \mathcal{Q}^{i+j+m}$ *is defined by:* $\forall q = (c, (b)_m, (A)_{m^2}, c_\pm, c_+, c_-) \in \mathcal{Q}^m$, $\operatorname{ext}_{i,j}(q) = (c, (b')_{i+j+m}, (A')_{(i+j+m)^2}, c_\pm, c_+, c_-) \in \mathcal{Q}^{i+j+m}$ *where* $b'_k = b_{k-i}$ *if* $i+1 \leq k \leq m+i$ *and 0 otherwise and* $A'_{k,l} = A_{k-i,l-i}$ *if* $i+1 \leq k, l \leq m+i$ *and 0 otherwise.*

Property 9.25 (Extension properties). *Let* $i, j \in \mathbb{N}$.

---

[4]Typically this involves a large number of loop unrolling, trying to minimize the number of actual uses of join/meet.

1. Let $t = (\epsilon, \epsilon_\pm, \epsilon_+, \epsilon_-) \in \mathbf{C}^m$ and $t' = (\epsilon', \epsilon_\pm, \epsilon_+, \epsilon_-) \in \mathbf{C}^{m+i+j}$ s.t. $\forall i+1 \leq k \leq m+i$, $\epsilon'_k = \epsilon_{k-i}$. Then $q(\epsilon', \epsilon_\pm, \epsilon_+, \epsilon_-) = \text{ext}_{i,j}(q)(\epsilon, \epsilon_\pm, \epsilon_+, \epsilon_-)$.
2. For all $q \in \mathcal{Q}^m$, $\gamma_\mathcal{Q}(q) = \gamma_\mathcal{Q}(\text{ext}_{i,j}(q))$.

Now, we can give a formal definition of the upper bound of two quadratic vectors.

Definition 9.26. *The upper bound $\sqcup_{\mathcal{Z}_\mathcal{Q}} : \mathcal{Z}_{\mathcal{Q}^m}^p \times \mathcal{Z}_{\mathcal{Q}^m}^p \to \mathcal{Z}_{\mathcal{Q}^{m+p}}^p$ is defined, for all $q = (c, b, A, c_\pm, c_+, c_-)$, $q' = (c', b', A', c'_\pm, c'_+, c'_-) \in \mathcal{Z}_{\mathcal{Q}^m}^p$ by:*

$$q \sqcup_{\mathcal{Z}_\mathcal{Q}} q' = (\text{ext}_{0,p}(q''_k))_{k \in [p]} + q^e \in \mathcal{Z}_{\mathcal{Q}^{m+p}}^p$$

where $q'' = (c'', (b'')_m^p, (A'')_{m^2}^p, c''^p_\pm, c''^p_+, c''^p_-) \in \mathcal{Z}_{\mathcal{Q}^m}^p$ with, for all $k \in [p]$:

- $(c'')_k = \text{mid}(\gamma_\mathcal{Q}(\pi_k(q)) \cup \gamma_\mathcal{Q}(\pi_k(q')))$;
- $\forall t \in \{\pm, +, -\}, c''_{t,k} = \text{argmin}(c_{t,k}, c'_{t,k})$;
- $\forall i \in [m], (b'')_{k,i} = \text{argmin}(b_{k,i}, b'_{k,i})$;
- $\forall i, j \in [m], (A'')_{k,i,j} = \text{argmin}(A_{k,i,j}, A'_{k,i,j})$;

and $\forall k \in [p], q^e_k = \text{ext}_{(m+k-1),(p-k)}(\alpha_\mathcal{Q}(C_k \sqcup_\mathcal{I} C'_k))$ with $C_k = \gamma_\mathcal{Q}(\pi_k(q) - \pi_k(q''))$ and $C'_k = \gamma_\mathcal{Q}(\pi_k(q') - \pi_k(q''))$.

Let us denote the Minkowski sum and the Cartesian product of sets, respectively, by $D_1 \oplus D_2 = \{d_1 + d_2 \mid d_1 \in D_1, d_2 \in D_2\}$ and $\prod_i^n D_i = \{(d_1, \dots, d_n) \mid \forall i \in [n], d_i \in D_i\}$. We have the nice characterization of the concretization of the upper bound given by Lemma 9.27.

Lemma 9.27. *By construction of q'' and q^e previously defined:*

$$\gamma_{\mathcal{Z}_\mathcal{Q}}\left((\text{ext}_{0,p}(q''_k))_{k \in [p]} + q^e\right) =$$

$$\gamma_{\mathcal{Z}_\mathcal{Q}}(q'') \oplus \prod_{k=1}^p \gamma_{\mathcal{Q}^{m+p}}(q^e_k).$$

Now, we state in Theorem 9.28 that the $\sqcup_{\mathcal{Z}_\mathcal{Q}}$ operator computes an upper bound of its operands with respect to the preorder $\sqsubseteq_{\mathcal{Z}_\mathcal{Q}}$.

Theorem 9.28. *For all $q, q' \in \mathcal{Z}_{\mathcal{Q}^m}^p$, $q \sqsubseteq_{\mathcal{Z}_\mathcal{Q}} q \sqcup_{\mathcal{Z}_\mathcal{Q}} q'$ and $q' \sqsubseteq_{\mathcal{Z}_\mathcal{Q}} q \sqcup_{\mathcal{Z}_\mathcal{Q}} q'$.*

Example 9.29. *Let Q and Q' be two quadratic vectors:*

$$Q = \begin{cases} x = -1 + \epsilon_1 - \epsilon_2 - \epsilon_{1,1} \\ y = 1 + 2\epsilon_2 + \epsilon_{1,2} \end{cases}$$

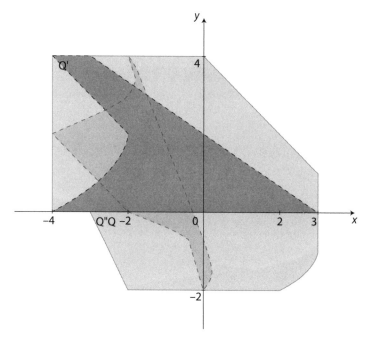

Figure 9.5. Upper bound computation.

$$Q' = \begin{cases} x = -2\epsilon_2 - \epsilon_{1,1} + \epsilon_+ \\ y = 1 + \epsilon_1 + \epsilon_2 + \epsilon_{1,2}. \end{cases}$$

The resulting quadratic vector $Q'' = Q \sqcup_{Z_Q} Q'$ is

$$Q'' = \begin{cases} x = -\epsilon_2 - \epsilon_{1,1} + 2\epsilon_3 \\ y = 1 + \epsilon_2 + \epsilon_{1,2} + \epsilon_4. \end{cases}$$

TRANSFER FUNCTIONS

The two operators `guard` and `assign` over the expressions *RelExpr* and *Expr* are defined like in a nonrelational abstract domain, as described in [49, §2.4.4]. Each operator relies on the forward semantics of numerical expressions, computed within arithmetics operators in $\mathcal{Q}$:

Definition 9.30. *Let $\mathcal{V}$ be a finite set of variables. Let $[\![\cdot]\!]_{\mathcal{Q}}(\mathcal{V} \to \mathcal{Q}) \to \mathcal{Q}$ be the semantics evaluation of an expression in an environment mapping variables to quadratic forms.*

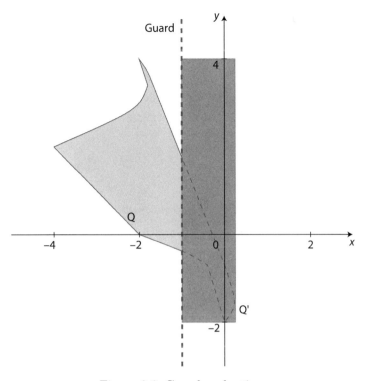

Figure 9.6. Guard evaluation.

$$
[\![v]\!]_{\mathcal{Q}}(Env) \quad = \quad \begin{array}{l} \pi_k(Env) \ where \ k \in [p] \\ is \ the \ index \ of \ v \in \mathcal{V} \ in \ Env \end{array}
$$

$$
[\![e_1 \ bop \ e_2]\!]_{\mathcal{Q}}(Env) \quad = \quad \begin{array}{l} [\![e_1]\!]_{\mathcal{Q}}(Env) \\ bop_{\mathcal{Q}} \ [\![e_2]\!]_{\mathcal{Q}}(Env) \end{array}
$$

$$
[\![uop \ e]\!]_{\mathcal{Q}}(Env) \quad = \quad uop_{\mathcal{Q}}[\![e]\!]_{\mathcal{Q}}(Env)
$$

Guards, i.e., tests, are enforced through the classical combination of forward and backward operators. Backward operators are the usual fallback operators, e.g., $[\![x + y]\!]^{\leftarrow} = (x \sqcap_{\mathcal{Q}} ([\![x + y]\!] -_{\mathcal{Q}} y), y \sqcap_{\mathcal{Q}} ([\![x + y]\!] -_{\mathcal{Q}} x))$ where $\sqcap_{\mathcal{Q}}$ denotes the meet of quadratic forms. As for upper bound computation, no best lower bound exists and such meet operator in $\mathcal{Q}$ has to compute a safe but imprecise upper bound of maximal lower bounds.

The meet over $\mathcal{Q}^m$ works by (i) projecting each argument to intervals using $\gamma_{\mathcal{Q}}$; (ii) performs the meet computation, (iii) reinjects the resulting closed bounded interval to $\mathcal{Q}$ using $\alpha_{\mathcal{Q}}$, and (iv) introduces it through a fresh noise symbol.

The meet over $\mathcal{Z}_{\mathcal{Q}^m}^p$ is defined as the lift of $\mathcal{Q}^m$ meet to quadratic vectors. Formally:

Definition 9.31. *The meet* $\sqcap_Q : Q^m \times Q^m \to Q^1$ *is defined by:*

$$\forall x, y \in Q^m, x \sqcap_Q y \triangleq \alpha_Q (\gamma_Q(x) \sqcap_\mathcal{I} \gamma_Q(y)).$$

The meet $\sqcap_{Z_Q} : \mathcal{Z}^p_{Q^m} \times \mathcal{Z}^p_{Q^m} \to \mathcal{Z}^p_{Q^p}$ *is defined, for all* $x, y \in \mathcal{Z}^p_{Q^m}$, *by* $z = x \sqcap_{Z_Q} y \in \mathcal{Z}^p_{Q^p}$ *where:*

$$\forall i \in [p], z_i = \pi_i(x) \sqcap_Q \pi_i(y)$$

when $\pi_i(x) \neq \pi_i(y), \pi_i(x)$ *otherwise.*

Example 9.32. *Let Q be the following quadratic vector. The meet with the constraint $x + 1 \geq 0$ produces the resulting quadratic vector Q':*

$$Q = \begin{cases} x = -1 + \epsilon_1 - \epsilon_2 - \epsilon_{1,1} \\ y = 1 + 2\epsilon_2 + \epsilon_{1,2} \end{cases}$$

$$Q' = \begin{cases} x = -\frac{3}{8} + \frac{5}{8}\epsilon_3 \\ y = 1 + 2\epsilon_2 + \epsilon_{1,2}. \end{cases}$$

9.3.3.5 *Floating-point computations*

In the specific case of quadratic forms, the term in $\epsilon_\pm$ is used to accumulate floating-point errors: the number of error terms does not increase due to floating-point computation. The generalization to zonotopes is straightforward since numerical operations are evaluated at form level.

We illustrate here these principles on the addition and external multiplication operators. All initial arithmetic operations are provided in Messine and Touhami [61]. Here, we gather the additive and multiplicative errors of each operator and accumulate them in $\epsilon_\pm$ terms, following [29] methodology.

1. Addition.

   We consider the addition of two quadratic forms $x = (x_0, (x_i), (x_{ij}), x_\pm, x_+, x_-)$ and $y = (y_0, (y_i), (y_{ij}), y_\pm, y_+, y_-)$. The addition of x and y is modified to consider these generated errors:

   $$(x_0, (x_i), (x_{ij}), x_\pm, x_+, x_-)$$
   $$+_Q (y_0, (y_i), (y_{ij}), y_\pm, y_+, y_-) =$$
   $$\begin{pmatrix} x_0 + y_0, (x_i + y_i), (x_{ij} + y_{ij}), \\ x_\pm + y_\pm + r\_err, x_+ + y_+, x_- + y_- \end{pmatrix}$$

   where

Table 9.1. Stolfi example [21] and Householder numerical results.

	Intervals		Affine Z.		Quad. Z.	
	val	*ms*	*val*	*ms*	*val*	*ms*
stolfi1	$[0., \infty]$	6	$[-\infty, +\infty]$	0	$[-0.85, 3.25]$	7
stolfi2	$[0., 3.60]$	10	$[-\infty, +\infty]$	7	$[-\infty, +\infty]$	4
stolfi3	$[0., 5.38]$	4	$[-2.98, 10.81]$	5	$[-0.62, 3.24]$	5
stolfi4	$[0., 2.23]$	10	$[-\infty, +\infty]$	6	$[-0.33, 3.26]$	11
stolfi5	$[0., 2.18]$	9	$[-0.42, 3.03]$	3	$[0.08, 2.38]$	11
stolfi6	$[0., 2.18]$	7	$[-0.71, 2.91]$	5	$[0.11, 2.30]$	6
stolfi7	$[0., 1.89]$	7	$[0.19, 2.23]$	3	$[0.29, 1.97]$	10
stolfi30	$[0.35, 1.43]$	15	$[0.48, 1.44]$	20	$[0.48, 1.43]$	18
stolfi40	$[0.40, 1.40]$	15	$[0.49, 1.40]$	20	$[0.50, 1.40]$	18
stolfi50	$[0.43, 1.38]$	19	$[0.50, 1.38]$	34	$[0.50, 1.38]$	21
stolfi55	$[0.44, 1.37]$	24	$[0.51, 1.37]$	33	$[0.51, 1.37]$	35
stolfi100	$[0.48, 1.34]$	29	$[0.52, 1.34]$	80	$[0.52, 1.34]$	66
stolfi200	$[0.51, 1.32]$	48	$[0.53, 1.32]$	337	$[0.53, 1.32]$	269
stolfi300	$[0.52, 1.31]$	73	$[0.53, 1.31]$	916	$[0.53, 1.31]$	554
stolfi400	$[0.52, 1.31]$	91	$[0.53, 1.31]$	1746	$[0.53, 1.31]$	1086
householder #3	$[0.21, 0.24]$	3	$[0.21, 0.24]$	9	$[0.21, 0.24]$	4
householder #4	$[0.17, 0.29]$	0	$[0.22, 0.25]$	7	$[0.22, 0.24]$	8
householder #5	$[0.03, 0.42]$	3	$[0.22, 0.25]$	8	$[0.22, 0.24]$	10
householder #6	$[-0.90, 1.66]$	3	$[0.22, 0.25]$	19	$[0.22, 0.24]$	14
householder #7	$[-1117.82, 1899.48]$	4	$[0.22, 0.25]$	27	$[0.22, 0.24]$	11
householder #8	$[-2.18e^{+18}, 3.70e^{+18}]$	5	$[0.22, 0.25]$	29	$[0.22, 0.24]$	11

Best method is highlighted. Results are shown with two decimal digit precision.

- $r\_err = max(|\uparrow_{+\infty}(err)|, |\uparrow_{-\infty}(err)|)$
- $err = \sum_{i,j=1}^{n} e^+(x_{ij}, y_{ij}) + \sum_{i=0}^{n} e^+(x_i, y_i) + e^+(x_\pm, y_\pm) + e^+(x_+, y_+) + e^+(x_-, y_-).$

2. External multiplication.

The operator $*_Q$ is modified to account for multiplicative errors:

$$\lambda *_Q (x_0, (x_i), (x_{ij}), x_\pm, x_+, x_-) =$$
$$(\lambda x_0, \lambda(x_i), \lambda(x_{ij}), |\lambda| x_\pm + r\_err, |\lambda| x_+, |\lambda| x_-)$$

where

- $r\_err = max(|\uparrow_{-\infty} (err)|, |\uparrow_{+\infty} (err)|)$.
- $err = \sum_{i=1}^{n} e^\times(\lambda, x_i) + \sum_{i,j=1}^{n} e^\times(\lambda, x_{ij}) + e^\times(\lambda, x_\pm) + e^\times(\lambda, x_-) + e^\times(\lambda, x_+)$
  $+$eta.

Table 9.1 presents numerical results, comparing analyses.

9.4   RELATED WORKS

Analysis of accuracy of finite-precision arithmetic is not a new topic. Multiple numerical methods have been proposed to address this issue: interval arithmetics, stochastic arithmetics, automatic differentiation, or error series. See [38, 57] for comparative surveys. In static analysis, most analyses rely on safely rounded intervals or on zonotopes [85, 94, 113]. These affine arithmetics based domains are the core of the tool Fluctuat [121] which is able to bound the numerical errors and identify the terms or variables that participate the most to the final error. These domains are also now implemented as an APRON [88] domain.

Chapter Ten

Convex Optimization and Numerical Issues

IN THIS CHAPTER we aim at providing the intuition behind convex optimization algorithms and address their effective use with floating-point implementation. The first section presents briefly the algorithms, assuming a real semantics, while Section 10.2 presents our approaches to obtain sound results.

10.1 CONVEX OPTIMIZATION ALGORITHMS

As outlined in Section 4.2.1 convex conic programming is supported by different methods depending on the cone considered.

The most known approach for linear constraints is the simplex method by Dantzig. While having an exponential-time complexity with respect to the number of constraints, the simplex method performs well in general. Intuitively, starting from a vertex of the convex polytope, it follows the hyperplan minimizing the cost of the objective function to reach the best neighbor vertex. It can be seen as a discrete and combinatorial method: it enumerates faces of the polytope but each step could be solved exactly, for example, with rational arithmetics; and each intermediate step characterizes exactly a vertex of the polytope. Theoretically the optimal solution is always on a vertex. Either the optimal solution is a unique vertex, or it corresponds to a constraint defining the feasible region. In that latter case, the optimal solution is defined as an infinite set of points of the hyperplan.

If the simplex method behaves properly, one of the advantages of its use is the obtention of a feasible optimal solution.

On the negative aspects, its complexity could diverge in some ill-shaped cases and it is not extensible to the larger set of linear problems over convex cones.

Another method is the set of interior point methods, initially proposed by Karmarkar [12] and made popular by Nesterov and Nemirovski [14, 15, 24]. They can be characterized as path-following methods in which a sequence of local linear problems are solved, typically by Newton's method.

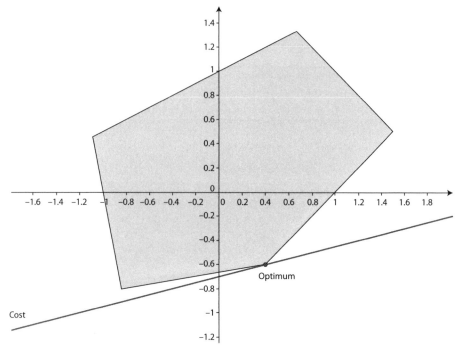

Figure 10.1. Linear optimization over a polytope.

10.1.1   Convex optimization with interior point method algorithms

An interior point method optimization is performed in two steps: A first one computes the analytical center of the convex set of constraints. This element is characterized by a logarithmic barrier function.[1]

Definition 10.1 (Analytical Center).   *Let $\bigwedge_i (f_i(x) \leq 0)$ be a set of convex inequalities, the analytical center is defined as the optimal solution of the convex problem*

$$minimize \ \phi(x) \ where \ \phi(x) = -\sum_i log(-f_i(x)). \qquad (10.1)$$

The computation of such value by solving the gradient equal to zero:

$$\nabla \phi(x) = \sum_i \frac{1}{-f_i(x)} \nabla f_i(x). \qquad (10.2)$$

---

[1]In order to keep the presentation simple we provide definition and illustration here on linear constraints instead of more general linear matrix inequalities.

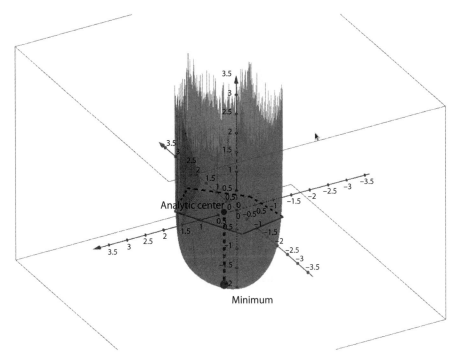

Figure 10.2. Barrier function characterizing the analytical center.

Intuitively this function tends to $+\infty$ when the x is near one of the linear constraints $f_i(x) = 0$. In case of linear constraints, $f_i(x) = a_i(x) - b_i$ and the expression of $\nabla \phi$ becomes $\sum_i \frac{1}{-f_i(x)} a_i$. The following figure illustrates such barrier function on the preceding polytope.

Path-following algorithms encode the search of the solution of a linear optimizing problem as a sequence of local Newton problems. Phase I steps compute the starting point, the analytical center of the set of constraints. Then Phase II performs a sequence of local Newton steps. It characterizes a notion of central path, a function $\widetilde{f}(t; x)$ that integrates both the constraint $\phi(x)$ to be in the interior of the set of linear constraints, and the (linear) objective function $f(x)$:

Definition 10.2 (Central Path).

$$\widetilde{f}(t; x) = t \times f(x) + \phi(x)$$

Note that when $t = 0$ we have the analytical center $\widetilde{f}(0; x) = \phi(x)$, and when $t \to +\infty$, this constraint vanishes while the objective function becomes stronger in the expression.

Without arguing too much about the details of the computation, let us illustrate on this simple example the computed step towards the optimal solution:

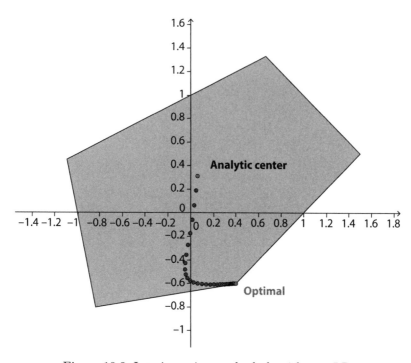

Figure 10.3. Interior point method algorithm in LP.

Interior point methods compute a sequence of intermediate feasible solutions. Each step is performed by computing a linear direction and a step length, such that the next value remains in the neighborhood of the central path while improving the objective function cost. This sequence stops when a required precision is reached. On a typical implementation the stopping criteria is around 10^{-7}.

In contrast to the simplex method, we see clearly that this method will never compute the exact solution but is able to approximate it as precisely as required. On the good sides, interior points methods can be defined for more general settings than linear constraints, for example, on the large set of convex conic sets (quadratic programming, second-order conic programming, semidefinite programming, etc.). On the complexity side, the method has a polynomial complexity in the number of variables. Typical uses in software work on thousands of variables while efficient uses can scale to hundreds of thousands of variables when smartly implemented, e.g., on FPGA [135].

10.1.2   Primal-dual feasibility

Recall the definition of duality in optimization (Section 4.2.3). Among the most efficient implementations of interior point methods, primal-dual methods rely on the notion of duality gap to measure the distance to the optimal solution.

When manipulating matrix variables, the inequality (4.28) can be rephrased on the Hilbert space of positive definite matrices:

$$
\begin{array}{llll}
d^* = & \max_X \ \langle c, X \rangle & \leq \quad p^* = & \min_y \ -\langle y, b \rangle \\
& \text{s.t.} \quad AX = b & & \text{s.t.} \quad -c - A'y = Z \\
& \qquad X \succeq 0 & & \qquad Z \succeq 0
\end{array}
\tag{10.3}
$$

where d^* and p^* denote, respectively, the dual and primal solutions.

The distance between two feasible points of primal and dual problems is called the duality gap:

$$\langle c, X \rangle + \langle y, b \rangle.$$

The barrier function of the method is then characterized using the central path expression as the combination of the analytical center of both primal and dual problems, and the duality gap as the objective function to minimize.

However, it may happen that either the primal or the dual problem are ill-defined: they can be unfeasible or unbounded, i.e., $d^* = \infty$ or $p^* = -\infty$. According to duality theorem, if both primal and dual solutions are finite, then both problems are feasible. Otherwise, either the primal solution is unbounded and its dual unfeasible, or the dual solution is unbounded and the primal unfeasible.

10.1.3   Issues

These convex optimization methods have now reached sufficient maturity and are now used in a large set of contexts. However, for their specific use in formal verification, one needs to cope with the following issues:

1. In contrast to simplex methods, these methods never reach the optimal value. However, they admit an exponential convergence and could reach an arbitrary precise solution. On the positive side, the computed suboptimal solution, as well as all intermediate iterates of Phase II, are feasible solutions of the set of constraints. Therefore, when the objective function is not crucial–as it is when we synthesize our invariants with these techniques–these solutions can be used safely.
2. Because of floating-point errors, the actual ϵ-optimal solution may be (slightly) an unfeasible solution. This is particularly difficult when manipulating equality constraints.

While often neglected, this second issue has to be addressed in our formal framework. This is of great importance since we rely on computed solutions to bound the behavior of the analyzed systems. In case of SOS programming, the underlying SDP matrix is associated to a set of equality constraints enabling the characterizing of the positive polynomial. The reconstruction of such solution polynomial with floating-point arithmetics has also to be addressed in a formal way.

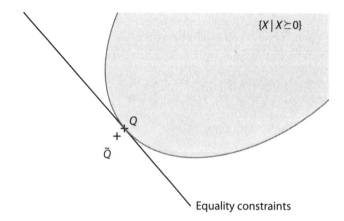

Figure 10.4. Floating-point errors with interior point methods.

10.2 GUARANTEED FEASIBLE SOLUTIONS WITH FLOATS

As we saw in the preceding figures, the optimal value of a linear objective function with respect to a set of convex constraints is always on the frontier of the convex set. In SDP optimization this could be mapped to the search of the point Q within the ellipsoid characterized by the set of constraints $X \succeq 0$, i.e., $\forall x \in \mathbb{R}, x^\mathsf{T} X x \geq 0$.

Because of floating-point errors, the computed value could be slightly outside of the set of constraints, giving an unfeasible, and therefore unsound, solution $\widetilde{Q}$.

10.2.1 Computation over a constrained cone

Since these path-following algorithms never actually reach the optimal value but stay within the interior of the set of constraints, the stopping criteria depends on the achievement of a given precision, typically a duality gap of 10^{-7}. Without any formal result, remark that, for all $|x| < 10^p$, $|x^2| < 10^{2p}$, so when computing with a precision of 16 digits, i.e., $\mathrm{ulp}(1) = 10^{-16}$, half the digits are lost by the quadratic form, leading to an imprecision of $\sqrt{(10^{16})} = 10^{-8}$ for Q.

A heuristic to ensure remaining within the interior of the set of constraints is to pad the convex constraint by this precision of 10^{-8}. Figure 10.5 presents such a constrained set of conic constraints. One can ensure that the floating-point value $\widetilde{Q}$ still remains within the convex set.

This padding may fail when the set of constraints admit an empty interior, which cannot be padded. This happens typically for our invariant search when the dynamic considered admits fixpoints; Let us consider the dynamics f and x_{fp} such as fixpoint; our search for an inductive invariant v will lead to a constraint of the form:

$$v \circ f - v \leq 0.$$

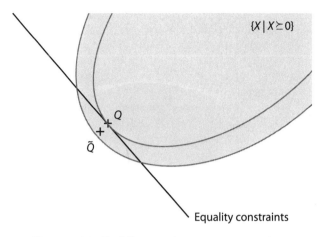

Figure 10.5. Padding conic convex constraints.

But since $f(x_{fp}) = x_{fp}$, we have

$$v(f(x_{fp})) - v(x_{fp}) = v(x_{fp}) - v(x_{fp}) = 0$$

and this problem admits an empty interior since v should be exactly equal to zero at x_{fp}. In this case the previous technique over-approximating the floating point semantics of f will impose an even more difficult constraint:

$$
\begin{aligned}
&\ v(f(x_{fp})) - v(x_{fp}) - err_{f(x_{fp})} \\
=&\ v(x_{fp}) - v(x_{fp}) - err_{f(x_{fp})} \\
=&\ -err_{f(x_{fp})} = 0
\end{aligned}
$$

with $err_{f(x_{fp})} > 0$.

In the linear case, when considering a stable system, there is a single fixpoint which is often zero. A first solution is to impose the Lyapunov function v to admit a zero coefficient for the monomial 1, allowing the error to vanish in zero.

For the more general setting, a solution is to encode inductiveness differently as a convex constraint: instead of $v \circ f - v \leq 0$ one can rely on $\alpha\, v \circ f - v \leq 0$ with $\alpha > 1$. This would correspond to a relative padding, applying a growth factor to the semialgebraic set defined by $v \leq 0$.

Considering, in addition, the floating-point semantics of the analyzed program, our inductiveness constraint in the polynomial case becomes:

$$p - \alpha\, p \circ T^i \sigma^i + \sum_{j=1}^{n_i} \mu_j^i r_j^i - err_{p \circ T^i} \geq 0$$

where $err_{p \circ T^i}$ is an upper bound on the floating-point errors obtained when evaluating $p \circ T^i$.

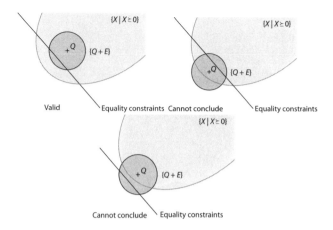

Figure 10.6. Sound positivity check using SOS.

10.2.2   Validate feasibility of the solution

Despite the additional padding, it may happen that the computed property is not strictly verifying the constraints. It then remains to formally check that the computed values satisfy the set of positive constraints.

When manipulating directly LMIs, this can be done by computing the matrix associated to each positive constraint, for instance, with exact rational arithmetic, and then checking that the result is positive definite. This last check can be performed using a Cholesky decomposition. For the sake of efficiency, this decomposition can itself be performed using floating-point arithmetic by carefully bounding the rounding errors [64]. The resulting algorithm being nontrivial, it has been proved using the proof assistant Coq [144].

When manipulating SOS polynomials, additional equality constraints relate matrix coefficients to the coefficients of the SOS polynomial. It remains to check that the floating-point polynomial solution is such that it verifies all positive constraints.

To show that a polynomial constraint e is positive when all its unknown variables have been valued by the optimization solver, we evaluate in rational arithmetics the polynomial p associated to the expression e. SOS optimization will then exhibit a floating-point matrix Q such that $p = x^\mathsf{T} Q x$.

Because of numerical imprecision, due both to the method (interior point) and the floating-point rounding errors, we rather have $p = x^\mathsf{T} Q x + x^\mathsf{T} E x = x^\mathsf{T} (Q + E) x$ where E denotes the error produced during the SOS programming.

This error term $x^\mathsf{T} E x$ can be characterized as $p - x^\mathsf{T} Q x$, and we can identify an upper bound ϵ on the error per coefficient:

$$\forall i, j, |E|i, j \le \epsilon.$$

In order to prove that there exists a positive semidefinite matrix $Q + E \succeq 0$, we check the stronger condition $Q - n\epsilon Id \succeq 0$ where n is the dimension of Q. The Figure 10.6 presents the different results: Q is not exactly on the constraints but in a neighborhood. In the first case, all the matrices $Q - n\epsilon Id$ are positive definite, so there exists a witness of positivity for the constraint p. In the second case, the constraint intersects the cone and p is positive, but the imprecise computed Q does not enable us to prove that $Q - n\epsilon Id \succeq 0$. In the last case, p is not positive, but Q was produced as an SOS witness of positivity for p. Thanks to the invalidity of the check, we are able to reject this value.

Bibliography

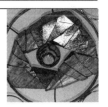

[1] Théodore Samuel Motzkin. "Two consequences of the transposition theorem on linear inequalities." In: *Econometrica* 19.2 (1951), pp. 184–185.

[2] Richard FitzHugh. "Impulses and physiological states in theoretical models of nerve membrane." In: *Biophys J.* 1 (1961), pp. 445–466. ISSN: 0006-3495.

[3] Robert W. Floyd. "Assigning meanings to programs." In: *Proceedings of Symposium on Applied Mathematics* 19 (1967), pp. 19–32.

[4] Charles A. R. Hoare. "An axiomatic basis for computer programming." In: *Commun. ACM* 12 (Oct. 1969), pp. 576–580.

[5] Jan C. Willems. "Dissipative dynamical systems part I: General theory." In: *Archive for Rational Mechanics and Analysis* 45.5 (1972), pp. 321–351.

[6] Edsger Wybe Dijkstra. *A Discipline of Programming.* Prentice-Hall, 1976.

[7] Michael Karr. "Affine relationships among variables of a program." In: *Acta Inf.* 6 (1976), pp. 133–151. DOI: 10.1007/BF00268497.

[8] Patrick Cousot and Radhia Cousot. "Abstract interpretation: A unified lattice model for static analysis of programs by construction or approximation of fixpoints." In: *Proceedings of the 4th ACM SIGACT-SIGPLAN Symposium on Principles of Programming Languages.* POPL '77, New York, NY, USA: ACM, 1977, pp. 238–252.

[9] Patrick Cousot and Radhia Cousot. "Systematic design of program analysis frameworks." In: *Conference Record of the Sixth Annual ACM SIGPLAN-SIGACT Symposium on Principles of Programming Languages.* San Antonio, Texas: ACM Press, New York, NY, 1979, pp. 269–282.

[10] D. H. Martin and D. H. Jacobson. "Copositive matrices and definiteness of quadratic forms subject to homogeneous linear inequality constraints." In: *Linear Algebra and Its Applications* 35 (1981), pp. 227–258. ISSN: 0024-3795. DOI: dx.doi.org/10.1016/0024-3795(81)90276-7.

[11] John C. Doyle. "Analysis of feedback systems with structured uncertainties." In: *IEE Proceedings D-Control Theory and Applications* 129.6 (Nov. 1982), pp. 242–250. ISSN: 0143-7054. DOI: 10.1049/ip-d.1982.0053.

[12] Narendra Karmarkar. "A new polynomial-time algorithm for linear programming." In: *Combinatorica* 4.4 (Dec. 1984), pp. 373–395.

[13] Paul Caspi, Daniel Pilaud, Nicolas Halbwachs, and John Plaice. "Lustre: A declarative language for programming synchronous systems." In: *POPL*. ACM Press, 1987, pp. 178–188. ISBN: 0-89791-215-2.

[14] Yurii Nesterov and Arkadi Nemirovski. "A general approach to the design of optimal methods for smooth convex functions minimization." In: *Ekonomika i Matem. Metody*, 24 (1988), pp. 509–517.

[15] Yurii Nesterov and Arkadi Nemirovski. *Self-Concordant Functions and Polynomial Time Methods in Convex Programming*. Materialy po matematicheskomu obespecheniiu EVM. USSR Academy of Sciences, Central Economic & Mathematic Institute, 1989.

[16] Gene F. Franklin, Michael L. Workman, and Dave Powell. *Digital Control of Dynamic Systems*. 2nd. Boston, MA, USA: Addison-Wesley Longman Publishing Co., Inc., 1990. ISBN: 0201820544.

[17] Stephen A. Vavasis. "Quadratic programming is in NP." In: *Information Processing Letters* 36.2 (1990), pp. 73–77. ISSN: 0020-0190. DOI: dx.doi.org/10.1016/0020-0190(90)90100-C.

[18] Nicolas Halbwachs, Paul Caspi, Pascal Raymond, and Daniel Pilaud. "The synchronous data-flow programming language LUSTRE." In: *Proceedings of the IEEE* 79.9 (Sept. 1991), pp. 1305–1320.

[19] Patrick Cousot and Radhia Cousot. "Abstract interpretation frameworks." In: *Journal of Logic and Computation* 2.4 (Aug. 1992), pp. 511–547.

[20] Qinping Yang. "Minimum decay rate of a family of dynamical systems." PhD thesis. Stanford University, 1992.

[21] Joao L. D. Comba and Jorge Stolfi. *Affine Arithmetic and Its Applications to Computer Graphics*. 1993.

[22] Nicolas Halbwachs, Fabienne Lagnier, and Pascal Raymond. "Synchronous observers and the verification of reactive systems." In: *AMAST*. Ed. by Maurice Nivat, Charles Rattray, Teodor Rus, and Giuseppe Scollo. Workshops in Computing. Springer, 1993, pp. 83–96. ISBN: 3-540-19852-0.

[23] Stephen Boyd, Laurent El Ghaoui, Éric Féron, and Venkataramanan Balakrishnan. *Linear Matrix Inequalities in System and Control Theory*. Vol. 15. SIAM. Philadelphia, PA: SIAM, June 1994. ISBN: 0-89871-334-X.

[24] Yurii Nesterov and Arkadi Nemirovski. *Interior-point Polynomial Algorithms in Convex Programming*. Vol. 13. Studies in Applied Mathematics. Society for Industrial and Applied Mathematics, 1994.

[25] Alexandre Megretski and Anders Rantzer. *System Analysis via Integral Quadratic Constraints – Part I*. 1995.

[26] Nicholas J. Higham. *Accuracy and Stability of Numerical Algorithms*. Philadelphia, PA, USA: Society for Industrial and Applied Mathematics, 1996. ISBN: 0898713552.

[27] William S. Levine. *The Control Handbook*. The electrical engineering handbook series. Boca Raton (Fl.): CRC Press New York, 1996. ISBN: 0-8493-8570-9.

[28] Lieven Vandenberghe and Stephen Boyd. "Semidefinite Programming." In: *SIAM Review* 38.1 (1996), pp. 49–95. DOI: 10.1137/1038003. eprint: epubs.siam.org/doi/pdf/10.1137/1038003.

[29] Jorge Stolfi and Luiz Henrique de Figueiredo. *Self-Validated Numerical Methods and Applications*. Brazilian Mathematics Colloquium monograph. IMPA, Rio de Janeiro, Brazil, July 1997.

[30] Brian Borchers. "CSDP, A C library for semidefinite programming." In: *Optimization Methods and Software* 11.1-4 (1999), pp. 613–623, DOI: 10.1080/10556789908805765. eprint: www.tandfonline.com/doi/pdf/10.1080/10556789908805765.

[31] Luigi Ambrosio, Nicola Fusco, and Diego Pallara. *Functions of Bounded Variation and Free Discontinuity Problems*. Oxford mathematical monographs. Autres tirages: 2006. Oxford, New York: Clarendon Press, 2000. ISBN: 0-19-850245-1.

[32] Erling D. Andersen and Knud D. Andersen. "The Mosek interior point optimizer for linear programming: An implementation of the homogeneous algorithm." English. In: *High Performance Optimization*. Ed. by Hans Frenk, Kees Roos, Tamás Terlaky, and Shuzhong Zhang. Vol. 33. Applied Optimization. Springer US, 2000, pp. 197–232. ISBN: 978-1-4419-4819-9. DOI: 10.1007/978-1-4757-3216-0_8.

[33] Keith Glover, Glenn Vinnicombe, and George Papageorgiou. "Guaranteed multi-loop stability margins and the gap metric." In: *CDC*. Vol. 4. IEEE. 2000, pp. 4084–4085.

[34] Saidhakim Dododzhanovich Ikramov and N.V. Savel'eva. "Conditionally definite matrices." In: *Journal of Mathematical Sciences* 98.1 (2000), pp. 1–50. ISSN: 1072-3374. DOI: 10.1007/BF02355379.

[35] Domenico Mignone, Giancarlo Ferrari-Trecate, and Manfred Morari. "Stability and stabilization of piecewise affine and hybrid systems: an LMI approach." In: *Decision and Control, 2000. Proc. of the 39th IEEE Conference*. Vol. 1. 2000, pp. 504–509. DOI: 10.1109/CDC.2000.912814.

[36] Mary Sheeran, Satnam Singh, and Gunnar Stålmarck. "Checking safety properties using induction and a SAT-solver." In: *FMCAD*. Ed. by Warren A. Hunt Jr. and Steven D. Johnson. Vol. 1954. LNCS. Springer, 2000, pp. 108–125. ISBN: 3-540-41219-0. DOI: 10.1007/3-540-40922-X_8.

[37] Vincent D. Blondel, Olivier Bournez, Pascal Koiran, Christos H. Papadimitriou, and John N. Tsitsiklis. "Deciding stability and mortality of piecewise affine dynamical systems." In: *Theoretical Computer Science A* 1–2.255 (2001), pp. 687–696.

[38] Eric Goubault. "Static analyses of the precision of floating-point operations." In: *Proceedings of the 8th International Symposium on Static Analysis*. SAS '01. London, UK: Springer-Verlag, 2001, pp. 234–259.

[39] Jean-Bernard Lasserre. "Global optimization with polynomials and the problem of moments." In: *SIAM Journal on Optimization* 11.3 (2001), pp. 796–817.

[40] Peter W. O'Hearn, John C. Reynolds, and Hongseok Yang. "Local reasoning about programs that alter data structures." In: *Computer Science Logic, 15th International Workshop, CSL 2001. 10th Annual Conference of the EACSL, Paris, France, September 10-13, 2001, Proceedings*. Ed. by Laurent Fribourg. Vol. 2142. Lecture Notes in Computer Science. Springer, 2001, pp. 1–19. ISBN: 3-540-42554-3. DOI: 10.1007/3-540-44802-0_1.

[41] Sam Owre and Natarajan Shankar. *Theory Interpretation in PVS*. Tech. rep. SRI International, 2001.

[42] Glenn Vinnicombe. *Uncertainty and Feedback: H [infinity] Loop-shaping and the [nu]-gap Metric*. World Scientific, 2001.

[43] Patrick Baudin, Anne Pacalet, Jacques Raguideau, Dominique Schoen, and Nicky Williams. "CAVEAT: A tool for software validation." In: *2002 International Conference on Dependable Systems and Networks (DSN 2002), 23-26 June 2002, Bethesda, MD, USA, Proceedings*. IEEE Computer Society, 2002, p. 537. ISBN: 0-7695-1597-5. DOI: 10.1109/DSN.2002.1028953.

[44] Markus Müller-Olm and Helmut Seidl. "Polynomial constants are decidable." In: *Static Analysis, 9th International Symposium, SAS 2002, Madrid, Spain, September 17-20, 2002, Proceedings*. Ed. by Manuel V. Hermenegildo and Germán Puebla. Vol. 2477. Lecture Notes in Computer Science. Springer, 2002, pp. 4–19. ISBN: 3-540-44235-9. DOI: 10.1007/3-540-45789-5_4.

[45] Pablo A. Parrilo. "Semidefinite programming relaxations for semialgebraic problems." English. In: *Mathematical Programming* 96.2 (2003), pp. 293–320. ISSN: 0025-5610. DOI: 10.1007/s10107-003-0387-5.

[46] Stephen Boyd and Lieven Vandenberghe. *Convex Optimization*. New York, NY, USA: Cambridge University Press, 2004.

[47] Jérôme Feret. "Static analysis of digital filters." In: *Programming Languages and Systems, 13th European Symposium on Programming, ESOP 2004, Held as Part of the Joint European Conferences on Theory and Practice of Software, ETAPS 2004, Barcelona, Spain, March 29 - April 2, 2004, Proceedings*. Ed. by David A. Schmidt. Vol. 2986. Lecture Notes in Computer Science. Springer, 2004, pp. 33–48. ISBN: 3-540-21313-9. DOI: 10.1007/978-3-540-24725-8_4.

[48] Johan Löfberg. "YALMIP: A toolbox for modeling and optimization in MATLAB." In: *Proceedings of the CACSD Conference*. Taipei, Taiwan, 2004.

[49] Antoine Miné. "Weakly relational numerical abstract domains." PhD thesis. École Polytechnique, Dec. 2004, p. 322.

[50] Sriram Sankaranarayanan, Henny Sipma, and Zohar Manna. "Non-linear loop invariant generation using Gröbner bases." In: *Proceedings of the 31st ACM SIGPLAN-SIGACT Symposium on Principles of Programming Languages, POPL 2004, Venice, Italy, January 14-16, 2004*. Ed. by Neil D. Jones and Xavier Leroy. ACM, 2004, pp. 318–329. ISBN: 1-58113-729-X. DOI: 10.1145/964001.964028.

[51] Pratik Biswas, Pascal Grieder, Johan Löfberg, and Manfred Morari. "A survey on stability analysis of discrete-time piecewise affine systems." In: *IFAC World Congress*. Prague, Czech Republic, July 2005.

[52] Alexandru Costan, Stéphane Gaubert, Eric Goubault, Matthieu Martel, and Sylvie Putot. "A policy iteration algorithm for computing fixed points in static analysis of programs." In: *Computer Aided Verification*. Ed. by Kousha Etessami and Sriram K. Rajamani. Vol. 3576. LNCS. Springer, 2005, pp. 462–475. ISBN: 3-540-27231-3.

[53] Patrick Cousot. "Proving program invariance and termination by parametric abstraction, Lagrangian relaxation and semidefinite programming."

English. In: *Verification, Model Checking, and Abstract Interpretation*. Ed. by Radhia Cousot. Vol. 3385. Lecture Notes in Computer Science. Springer Berlin Heidelberg, 2005, pp. 1–24. ISBN: 978-3-540-24297-0. DOI: `10.1007/978-3-540-30579-8_1`.

[54] Patrick Cousot, Radhia Cousot, Jérôme Feret, Laurent Mauborgne, Antoine Miné, David Monniaux, and Xavier Rival. "The ASTRÉE analyser." In: *ESOP*. Vol. 3444. LNCS. 2005, pp. 21–30,

[55] Jérôme Feret. "Analysis of mobile systems by abstract interpretation." PhD thesis. École polytechnique, Paris, France, 2005.

[56] Jérôme Feret. "Numerical abstract domains for digital filters." In: *International Workshop on Numerical and Symbolic Abstract Domains (NSAD)*. 2005.

[57] Matthieu Martel. "An overview of semantics for the validation of numerical programs." In: *Verification, Model Checking, and Abstract Interpretation, 6th International Conference, VMCAI 2005, Paris, France, January 17-19, 2005, Proceedings*. Ed. by Radhia Cousot. Vol. 3385. Lecture Notes in Computer Science. Springer, 2005, pp. 59–77. ISBN: 3-540-24297-X. DOI: `10.1007/978-3-540-30579-8_4`.

[58] Guy Norris and Mark Wagner. *Airbus A380: Superjumbo of the 21st Century*. Zenith Press, 2005. ISBN: 9780760322185.

[59] Stephen L. Campbell, Jean-Philippe Chancelier, and Ramine Nikoukhah. *Modeling and Simulation in Scilab, Scicos*. Springer, 2006.

[60] Bruno Dutertre and Leonardo de Moura. *The YICES SMT Solver*. Tech. rep. SRI International, 2006.

[61] Frédéric Messine and Ahmed Touhami. "A general reliable quadratic form: An extension of affine arithmetic." In: *Reliable Computing* 12.3 (2006), pp. 171–192.

[62] Antoine Miné. "The octagon abstract domain." In: *Higher-Order and Symbolic Computation* 19.1 (2006), pp. 31–100.

[63] Jorge Nocedal and Stephen J. Wright. *Numerical Optimization*. 2nd. New York: Springer, 2006.

[64] Siegfried M. Rump. "Verification of positive definiteness." In: *BIT Numerical Mathematics* 46 (2006), pp. 433–452.

[65] Clark Barrett and Cesare Tinelli. "CVC3." In: *Proceedings of the 19th International Conference on Computer Aided Verification (CAV '07)*. Ed. by Werner Damm and Holger Hermanns. Vol. 4590. Lecture Notes in

Computer Science. Berlin, Germany. Springer-Verlag, July 2007, pp. 298–302.

[66] Patrick Cousot, Radhia Cousot, Jérôme Feret, Laurent Mauborgne, Antoine Miné, David Monniaux, and Xavier Rival. "Combination of abstractions in the Astrée static analyzer." In: *Eleventh Annual Asian Computing Science Conference (ASIAN '06)*. Ed. by M. Okada and I. Satoh. Tokyo, Japan, LNCS 4435: Springer, Berlin, Dec. 2007, pp. 1–24.

[67] Jean-Christophe Filliâtre and Claude Marché. "The Why/Krakatoa/caduceus platform for deductive program verification." In: *Computer Aided Verification, 19th International Conference, CAV 2007, Berlin, Germany, July 3-7, 2007, Proceedings*. Ed. by Werner Damm and Holger Hermanns. Vol. 4590. Lecture Notes in Computer Science. Springer, 2007, pp. 173–177. ISBN: 978-3-540-73367-6. DOI: 10.1007/978-3-540-73368-3_21.

[68] Stéphane Gaubert, Eric Goubault, Ankur Taly, and Sarah Zennou. "Static analysis by policy iteration on relational domains." In: *ESOP*. Ed. by Racco De Nicola. Vol. 4421. LNCS. Springer, 2007, pp. 237–252. ISBN: 978-3-540-71314-2.

[69] Thomas Gawlitza and Helmut Seidl. "Precise fixpoint computation through strategy iteration." In: *ESOP*. Ed. by Racco De Nicola. Vol. 4421. LNCS. Springer, 2007, pp. 300–315. ISBN: 978-3-540-71314-2.

[70] Thomas Gawlitza and Helmut Seidl. "Precise relational invariants through strategy iteration." In: *CSL*. Ed. by Jacques Duparc and Thomas A. Henzinger. Vol. 4646. LNCS. Springer, 2007, pp. 23–40. ISBN: 978-3-540-74914-1.

[71] Ji-Woong Lee and Geir E. Dullerud. "Uniformly stabilizing sets of switching sequences for switched linear systems." In: *IEEE Trans. Automat. Contr.* 52.5 (2007), pp. 868–874. DOI: 10.1109/TAC.2007.895924.

[72] Ji-Woong Lee, Geir E. Dullerud, and Pramod P. Khargonekar. "An output regulation problem for switched linear systems in discrete time." In: *Proceedings of the 46th IEEE Conference on Decision and Control.* 2007, pp. 4993–4998.

[73] David Monniaux. "Applying the Z-transform for the static analysis of floating-point numerical filters." In: *CoRR* abs/0706.0252 (2007).

[74] Mathias Péron and Nicolas Halbwachs. "An abstract domain extending difference-bound matrices with disequality constraints." In: *Verification, Model Checking, and Abstract Interpretation, 8th International Conference, VMCAI 2007, Nice, France, January 14-16, 2007, Proceedings*. Ed.

by Byron Cook and Andreas Podelski. Vol. 4349. Lecture Notes in Computer Science. Springer, 2007, pp. 268–282. ISBN: 978-3-540-69735-0. DOI: 10.1007/978-3-540-69738-1_20.

[75] Andreas Podelski and Silke Wagner. "Region stability proofs for hybrid systems." In: *Formal Modeling and Analysis of Timed Systems, 5th International Conference, FORMATS 2007, Salzburg, Austria, October 3-5, 2007, Proceedings.* Ed. by Jean-François Raskin and P. S. Thiagarajan. Vol. 4763. Lecture Notes in Computer Science. Springer, 2007, pp. 320–335. ISBN: 978-3-540-75453-4. DOI: 10.1007/978-3-540-75454-1_23.

[76] Karl Johan Åström and Richard M. Murray. *Feedback Systems: An Introduction for Scientists and Engineers.* Princeton, NJ, USA: Princeton University Press, 2008. ISBN: 0691135762, 9780691135762.

[77] Dariusz Biernacki, Jean-Louis Colaço, Grégoire Hamon, and Marc Pouzet. "Clock-directed modular code generation for synchronous data-flow languages." In: *LCTES.* 2008.

[78] Sylvain Conchon, Evelyne Contejean, Johannes Kanig, and Stéphane Lescuyer. "CC(X): Semantic combination of congruence closure with solvable theories." In: *Electr. Notes Theor. Comput. Sci.* 198.2 (2008), pp. 51–69.

[79] Pierre-Loïc Garoche. "Static analysis of actors by abstract interpretation". PhD thesis. University of Toulouse, INPT, 2008.

[80] Wassim M. Haddad and VijaySekhar Chellaboina. *Nonlinear Dynamical Systems and Control: A Lyapunov-based Appr.* Princeton, NJ: Princeton University Press, 2008.

[81] David Monniaux. "The pitfalls of verifying floating-point computations." In: *ACM Trans. Program. Lang. Syst.* 30.3 (2008).

[82] Leonardo Mendonça de Moura and Nikolaj Bjørner. "Z3: An efficient SMT solver." In: *TACAS.* Ed. by C. R. Ramakrishnan and Jakob Rehof. Vol. 4963. Lecture Notes in Computer Science. Berlin, Heidelberg: Springer-Verlag, 2008, pp. 337–340. ISBN: 978-3-540-78799-0.

[83] Yannick Moy. "Sufficient preconditions for modular assertion checking." In: *Verification, Model Checking, and Abstract Interpretation, 9th International Conference, VMCAI 2008, San Francisco, USA, January 7-9, 2008, Proceedings.* Ed. by Francesco Logozzo, Doron A. Peled, and Lenore D. Zuck. Vol. 4905. Lecture Notes in Computer Science. Springer, 2008, pp. 188–202. ISBN: 978-3-540-78162-2. DOI: 10.1007/978-3-540-78163-9_18.

[84] Olivier Bouissou, Eric Goubault, Sylvie Putot, Karim Tekkal, and Franck Védrine. "Hybridfluctuat: A static analyzer of numerical programs within

a continuous environment." In: *Computer Aided Verification, 21st International Conference, CAV 2009, Grenoble, France, June 26 - July 2, 2009. Proceedings.* Vol. 5643. LNCS. 2009, pp. 620–626. ISBN: 978-3-642-02657-7. DOI: 10.1007/978-3-642-02658-4_46.

[85] Khalil Ghorbal, Eric Goubault, and Sylvie Putot. "The zonotope abstract domain Taylor1+." In: *CAV.* Vol. 5643. LNCS. 2009, pp. 627–633. ISBN: 978-3-642-02657-7.

[86] Eric Goubault and Sylvie Putot. "A zonotopic framework for functional abstractions." In: *CoRR* abs/0910.1763 (2009).

[87] Didier Henrion, Jean-Bernard Lasserre, and Carlo Savorgnan. "Approximate volume and integration for basic semialgebraic sets." In: *SIAM Review* 51.4 (2009), pp. 722–743. DOI: 10.1137/080730287. eprint: dx.doi.org/10.1137/080730287.

[88] Bertrand Jeannet and Antoine Miné. "Apron: A library of numerical abstract domains for static analysis." In: *CAV'09.* 2009, pp. 661–667, DOI: 10.1007/978-3-642-02658-4_52.

[89] Jean-Bernard Lasserre. *Moments, Positive Polynomials and Their Applications.* Imperial College Press optimization series. Imperial College Press, 2009. ISBN: 9781848164468.

[90] Saurabh Srivastava and Sumit Gulwani. "Program verification using templates over predicate abstraction." In: *SIGPLAN Not.* 44.6 (June 2009), pp. 223–234. ISSN: 0362-1340. DOI: 10.1145/1543135.1542501.

[91] Assalé Adjé, Stéphane Gaubert, and Eric Goubault. "Coupling policy iteration with semi-definite relaxation to compute accurate numerical invariants in static analysis." In: *ESOP.* Ed. by A. D. Gordon. Vol. 6012. Lecture Notes in Computer Science. Springer, 2010, pp. 23–42. ISBN: 978-3-642-11956-9.

[92] Éric Féron. "From control systems to control software." In: *Control Systems, IEEE* 30.6 (Dec. 2010), pp. 50 –71. ISSN: 1066-033X. DOI: 10.1109/MCS.2010.938196.

[93] Thomas Martin Gawlitza and Helmut Seidl. "Computing relaxed abstract semantics w.r.t. quadratic zones precisely." In: *SAS.* Ed. by Radhia Cousot and Matthieu Martel. Vol. 6337. LNCS. Springer, 2010, pp. 271–286. ISBN: 978-3-642-15768-4.

[94] Khalil Ghorbal, Eric Goubault, and Sylvie Putot. "A logical product approach to zonotope intersection." In: *CAV.* 2010, pp. 212–226.

[95] Stefan Ratschan and Zhikun She. "Providing a basin of attraction to a target region of polynomial systems by computation of Lyapunov-like functions." In: *SIAM J. Control and Optimization* 48.7 (2010), pp. 4377–4394. DOI: 10.1137/090749955.

[96] Siegfried M. Rump. "Verification methods: Rigorous results using floating-point arithmetic." In: *Acta Numerica* 19 (May 2010), pp. 287–449. ISSN: 1474-0508. DOI: 10.1017/S096249291000005X.

[97] Makoto Yamashita, Katsuki Fujisawa, Kazuhide Nakata, Maho Nakata, Mituhiro Fukuda, Kazuhiro Kobayashi, and Kazushige Goto. *A High-performance Software Package for Semidefinite Programs : SDPA7*. Tech. rep. Tokyo Institute of Technology, Tokyo, Japan: Dept. of Information Sciences, 2010.

[98] Andrew W. Appel. "Verified software toolchain-(invited talk)." In: *Programming Languages and Systems-20th European Symposium on Programming, ESOP 2011, Held as Part of the Joint European Conferences on Theory and Practice of Software, ETAPS 2011, Saarbrücken, Germany, March 26-April 3, 2011. Proceedings* Vol. 6602. LNCS. 2011, pp. 1–17. ISBN: 978-3-642-19717-8. DOI: 10.1007/978-3-642-19718-5_1.

[99] Michael Colón and Sriram Sankaranarayanan. "Generalizing the template polyhedral domain." In: *ESOP*. Vol. 6602. LNCS. 2011, pp. 176–195. ISBN: 978-3-642-19717-8.

[100] Eric Goubault and Sylvie Putot. "Static analysis of finite precision computations." In: *VMCAI*. Ed. by Ranjit Jhala and David A. Schmidt. Vol. 6538. LNCS. Springer, 2011, pp. 232–247. ISBN: 978-3-642-18274-7.

[101] Temesghen Kahsai, Yeting Ge, and Cesare Tinelli. "Instantiation-based invariant discovery." In: *NASA Formal Methods-Third International Symposium, NFM 2011, Pasadena, CA, USA, April 18-20, 2011. Proceedings*. 2011, pp. 192–206.

[102] Temesghen Kahsai and Cesare Tinelli. "Pkind: A parallel k-induction based model checker." In: *Proceedings 10th International Workshop on Parallel and Distributed Methods in verifiCation, PDMC 2011, Snowbird, Utah, USA, July 14, 2011*. Vol. 72. EPTCS. 2011, pp. 55–62. DOI: 10.4204/EPTCS.72.6.

[103] Ji-Woong Lee and Geir E. Dullerud. "Joint synthesis of switching and feedback for linear systems in discrete time." In: *Proceedings of the 14th ACM International Conference on Hybrid Systems: Computation and Control, HSCC 2011, Chicago, IL, USA, April 12-14, 2011*. Ed. by Marco Caccamo, Emilio Frazzoli, and Radu Grosu. ACM, 2011, pp. 201–210. ISBN: 978-1-4503-0629-4. DOI: 10.1145/1967701.1967731.

[104] Special C. RTCA. *DO-178C, Software Considerations in Airborne Systems and Equipment Certification.* 2011.

[105] Sriram Sankaranarayanan and Ashish Tiwari. "Relational abstractions for continuous and hybrid systems." In: *Computer Aided Verification-23rd International Conference, CAV 2011, Snowbird, UT, USA, July 14-20, 2011. Proceedings.* Ed. by Ganesh Gopalakrishnan and Shaz Qadeer. Vol. 6806. Lecture Notes in Computer Science. Springer, 2011, pp. 686–702. ISBN: 978-3-642-22109-5. DOI: 10.1007/978-3-642-22110-1_56.

[106] Mohamed Amin Ben Sassi, Romain Testylier, Thao Dang, and Antoine Girard. "Reachability analysis of polynomial systems using linear programming relaxations." In: *ATVA 2012* (2012). Ed. by Supratik Chakraborty and Madhavan Mukund, pp. 137–151. DOI: 10.1007/978-3-642-33386-6_12.

[107] Aaron Bradley. "Understanding IC3." In: *SAT 2012.* 2012, pp. 1–14.

[108] Pascal Cuoq, Florent Kirchner, Nikolai Kosmatov, Virgile Prevosto, Julien Signoles, and Boris Yakobowski. "Frama-C: A software analysis perspective." In: SEFM. Springer, 2012, pp. 233–247.

[109] Pierre-Loïc Garoche, Temesghen Kahsai, and Cesare Tinelli. *Invariant Stream Generators Using Automatic Abstract Transformers Based on a Decidable Logic.* 2012.

[110] Thomas Martin Gawlitza, Helmut Seidl, Assalé Adjé, Stéphane Gaubert, and Eric Goubault. "Abstract interpretation meets convex optimization." In: *J. Symb. Comput.* 47.12 (2012), pp. 1416–1446.

[111] Khalil Ghorbal, Franjo Ivančić, Gogul Balakrishnan, Naoto Maeda, and Aarti Gupta. "Donut domains: Efficient non-convex domains for abstract interpretation." In: *Verification, Model Checking, and Abstract Interpretation-13th International Conference, VMCAI 2012, Philadelphia, PA, USA, January 22-24, 2012. Proceedings.* Ed. by Viktor Kuncak and Andrey Rybalchenko. Vol. 7148. Lecture Notes in Computer Science. Springer, 2012, pp. 235–250. ISBN: 978-3-642-27939-3. DOI: 10.1007/978-3-642-27940-9_16.

[112] Eric Goubault, Tristan Le Gall, and Sylvie Putot. "An accurate join for zonotopes, preserving affine input/output relations." In: *ENTCS* 287 (2012), pp. 65–76.

[113] Eric Goubault, Sylvie Putot, and Franck Védrine. "Modular static analysis with zonotopes." In: *SAS.* Vol. 7460 LNCS. Springer, 2012, pp. 24–40. ISBN: 978-3-642-33124-4.

[114]   Heber Herencia-Zapana, Romain Jobredeaux, Sam Owre, Pierre-Loïc Garoche, Éric Féron, Gilberto Perez, and Pablo Ascariz. "PVS linear algebra libraries for verification of control software algorithms in C/ACSL." In: *NASA Formal Methods-Fourth International Symposium, NFM 2012, Norfolk, VA, USA, April 3-5, 2012. Proceedings.* Ed. by Alwyn Goodloe and Suzette Person. Vol. 7226. Lecture Notes in Computer Science. Springer, 2012, pp. 147–161. DOI: 10.1007/978-3-642-28891-3_15.

[115]   Milan Korda, Didier Henrion, and Colin Neil Jones. "Inner approximations of the region of attraction for polynomial dynamical systems." In: *ArXiv e-prints* (Oct. 2012). arXiv:1210.3184[math.OC].

[116]   Pierre Roux, Romain Jobredeaux, Pierre-Loïc Garoche, and Éric Féron. "A generic ellipsoid abstract domain for linear time invariant systems." In: *Proceedings of the 15th ACM International Conference on Hybrid Systems: Computation and Control, HSCC 2012, Beijing, China, April 17-19, 2012*, ed. by Thao Dang and Ian Mitchell. ACM, 2012, pp. 105–114. ISBN: 978-1-4503-1220-2. DOI: 10.1145/2185632.2185651.

[117]   John Rushby. "The versatile synchronous observer." In: *Proceedings of the 15th Brazilian Conference on Formal Methods: Foundations and Applications.* SBMF'12. Natal, Brazil: Springer-Verlag, 2012, p. 1. ISBN: 978-3-642-33295-1. DOI: 10.1007/978-3-642-33296-8_1.

[118]   Pierre-Loïc Garoche, Xavier Thirioux, and Temesghen Kahsai. *LustreC: A Modular Lustre Compiler.* 2012. URL: https://github.com/coco-team/lustrec.

[119]   Amir Ali Ahmadi and Raphael M. Jungers. "Switched stability of nonlinear systems via SOS-convex Lyapunov functions and semidefinite programming." In: *CDC '13*. 2013, pp. 727–732.

[120]   Jean-Charles Gilbert. *Éléments d'Optimisation Différentiable.* 2013.

[121]   Eric Goubault. "Static analysis by abstract interpretation of numerical programs and systems, and FLUCTUAT." In: *SAS* (2013), pp. 1–3.

[122]   Anvesh Komuravelli, Arie Gurfinkel, Sagar Chaki, and Edmund M. Clarke. "Automatic abstraction in SMT-based unbounded software model checking." In: *Computer Aided Verification-25th International Conference, CAV 2013, Saint Petersburg, Russia, July 13-19, 2013. Proceedings.* Ed. by Natasha Sharygina and Helmut Veith. Vol. 8044. Lecture Notes in Computer Science. Springer, 2013, pp. 846–862. ISBN: 978-3-642-39798-1. DOI: 10.1007/978-3-642-39799-8_59.

[123]   Milan Korda, Didier Henrion, and Colin Neil Jones. "Convex computation of the maximum controlled invariant set for discrete-time polynomial control systems." In *Decision and Control (CDC), 2013 IEEE 52nd*

Annual Conference on. Dec. 2013, pp. 7107–7112. DOI: 10.1109/CDC.2013.6761016.

[124] Milan Korda, Didier Henrion, and Colin Neil Jones. "Convex computation of the maximum controlled invariant set for polynomial control systems." In: *arXiv preprint arXiv:1303.6469* (2013).

[125] Eike Möhlmann and Oliver E. Theel. "Stabhyli: A tool for automatic stability verification of non-linear hybrid systems." In: *Proceedings of the 16th International Conference on Hybrid Systems: Computation and Control, HSCC 2013, April 8-11, 2013, Philadelphia, PA, USA*. Ed. by Calin Belta and Franjo Ivancic. ACM, 2013, pp. 107–112. ISBN: 978-1-4503-1567-8. DOI: 10.1145/2461328.2461347.

[126] Mehrdad Pakmehr, Timothy Wang, Romain Jobredeaux, Martin Vivies, and Éric Féron. "Verifiable control system development for gas turbine engines." In: *CoRR* abs/1311.1885 (2013).

[127] Pavithra Prabhakar and Miriam Garcia Soto. "Abstraction based model-checking of stability of hybrid systems." In: *Computer Aided Verification-25th International Conference, CAV 2013, Saint Petersburg, Russia, July 13-19, 2013. Proceedings*. Ed. by Natasha Sharygina and Helmut Veith. Vol. 8044. Lecture Notes in Computer Science. Springer, 2013, pp. 280–295. ISBN: 978-3-642-39798-1. DOI: 10.1007/978-3-642-39799-8_20.

[128] Pierre Roux. "Static analysis of control command systems: Synthetizing non-linear invariants." PhD thesis. Institut Supérieur de l'Aéronautique et de l'Espace, 2013.

[129] Pierre Roux and Pierre-Loïc Garoche. "Integrating policy iterations in abstract interpreters." In: *Automated Technology for Verification and Analysis-11th International Symposium, ATVA 2013, Hanoi, Vietnam, October 15-18, 2013. Proceedings*. Ed. by Dang Van Hung and Mizuhito Ogawa. Vol. 8172. Lecture Notes in Computer Science. Springer, 2013, pp. 240–254. ISBN: 978-3-319-02443-1. DOI: 10.1007/978-3-319-02444-8_18.

[130] Yassamine Seladji and Olivier Bouissou. "Numerical abstract domain using support functions." In: *NFM*. 2013.

[131] David Cachera, Thomas P. Jensen, Arnaud Jobin, and Florent Kirchner. "Inference of polynomial invariants for imperative programs: A farewell to Gröbner bases." In: *Sci. Comput. Program.* 93 (2014), pp. 89–109. DOI: 10.1016/j.scico.2014.02.028.

[132] Morgan Deters, Andrew Reynolds, Tim King, Clark W. Barrett, and Cesare Tinelli. "A tour of CVC4: How it works, and how to use it." In:

Formal Methods in Computer-Aided Design, FMCAD 2014, Lausanne, Switzerland, October 21-24, 2014. IEEE, 2014, p. 7. ISBN: 978-0-9835678-4-4. DOI: `10.1109/FMCAD.2014.6987586`.

[133]   Bruno Dutertre. "Yices 2.2." In: *Computer Aided Verification-26th International Conference, CAV 2014, Held as Part of the Vienna Summer of Logic, VSL 2014, Vienna, Austria, July 18-22, 2014. Proceedings.* Ed. by Armin Biere and Roderick Bloem. Vol. 8559. Lecture Notes in Computer Science. Springer, 2014, pp. 737–744. ISBN: 978-3-319-08866-2. DOI: `10.1007/978-3-319-08867-9_49`.

[134]   Didier Henrion and Milan Korda. "Convex computation of the region of attraction of polynomial control systems." In: *Automatic Control, IEEE Transactions on* 59.2 (2014), pp. 297–312. ISSN: 0018-9286. DOI: `10.1109/TAC.2013.2283095`.

[135]   Juan Luis Jerez, Paul J. Goulart, Stefan Richter, George A. Constantinides, Eric C. Kerrigan, and Manfred Morari. "Embedded online optimization for model predictive control at megahertz rates." In: *IEEE Trans. Automat. Contr.* 59.12 (2014), pp. 3238–3251. DOI: `10.1109/TAC.2014.2351991`.

[136]   Anvesh Komuravelli, Arie Gurfinkel, and Sagar Chaki. "SMT-based model checking for recursive programs." In: *Computer Aided Verification-26th International Conference, CAV 2014, Held as Part of the Vienna Summer of Logic, VSL 2014, Vienna, Austria, July 18-22, 2014. Proceedings.* Ed. by Armin Biere and Roderick Bloem. Vol. 8559. Lecture Notes in Computer Science. Springer, 2014, pp. 17–34. ISBN: 978-3-319-08866-2. DOI: `10.1007/978-3-319-08867-9_2`.

[137]   David Monniaux and Peter Schrammel. "Speeding up logico-numerical strategy iteration." In: *Static Analysis-21st International Symposium, SAS 2014, Munich, Germany, September 11-13, 2014. Proceedings.* Ed. by Markus Müller-Olm and Helmut Seidl. Vol. 8723. Lecture Notes in Computer Science. Springer, 2014, pp. 253–267. ISBN: 978-3-319-10935-0. DOI: `10.1007/978-3-319-10936-7_16`.

[138]   Assalé Adjé, Pierre-Loïc Garoche, and Alexis Werey. "Quadratic zonotopes-an extension of zonotopes to quadratic arithmetics." In: *Programming Languages and Systems-13th Asian Symposium, APLAS 2015, Pohang, South Korea, November 30-December 2, 2015, Proceedings.* 2015, pp. 127–145. DOI: `10.1007/978-3-319-26529-2_8`.

[139]   Xavier Allamigeon, Stéphane Gaubert, Eric Goubault, Sylvie Putot, and Nikolas Stott. "A scalable algebraic method to infer quadratic invariants of switched systems." In: *2015 International Conference on Embedded Software, EMSOFT 2015, Amsterdam, Netherlands, October 4-9, 2015.*

Ed. by Alain Girault and Nan Guan. IEEE, 2015, pp. 75–84. ISBN: 978-1-4673-8079-9. DOI: 10.1109/EMSOFT.2015.7318262.

[140] Adrien Champion, Rémi Delmas, and Michael Dierkes. "Generating property-directed potential invariants by quantifier elimination in a k-induction-based framework." In: *Sci. Comput. Program.* 103 (2015), pp. 71–87. DOI: 10.1016/j.scico.2014.10.004.

[141] Jean-Bernard Lasserre. "Tractable approximations of sets defined with quantifiers." English. In: *Mathematical Programming* 151.2 (2015), pp. 507–527. ISSN: 0025-5610. DOI: 10.1007/s10107-014-0838-1.

[142] Victor Magron, Didier Henrion, and Jean-Bernard Lasserre. "Semidefinite approximations of projections and polynomial images of semialgebraic sets." In: *SIAM Journal on Optimization* 25.4 (2015), pp. 2143–2164. DOI: 10.1137/140992047. eprint: dx.doi.org/10.1137/140992047.

[143] Mendes Oulamara and Arnaud J. Venet. "Abstract interpretation with higher-dimensional ellipsoids and conic extrapolation." In: *Computer Aided Verification-27th International Conference, CAV 2015, San Francisco, CA, USA, July 18-24, 2015, Proceedings, Part I.* Ed. by Daniel Kroening and Corina S. Pasareanu. Vol. 9206. Lecture Notes in Computer Science. Springer, 2015, pp. 415–430. ISBN: 978-3-319-21689-8. DOI: 10.1007/978-3-319-21690-4_24.

[144] Pierre Roux. "Formal proofs of rounding error bounds." English. In: *Journal of Automated Reasoning* (2015), pp. 1–22. ISSN: 0168-7433. DOI: 10.1007/s10817-015-9339-z.

[145] Timothy Wang. "Credible autocoding of control software." PhD thesis. School of Engineering–Georgia Tech, 2015.

[146] Egor George Karpenkov, David Monniaux, and Philipp Wendler. "Program analysis with local policy iteration." In: *Verification, Model Checking, and Abstract Interpretation-17th International Conference, VMCAI 2016, St. Petersburg, FL, USA, January 17-19, 2016. Proceedings.* Ed. by Barbara Jobstmann and K. Rustan M. Leino. Vol. 9583. Lecture Notes in Computer Science. Springer, 2016, pp. 127–146. ISBN: 978-3-662-49121-8. DOI: 10.1007/978-3-662-49122-5_6.

[147] Timothy Wang, Romain Jobredeaux, Heber Herencia-Zapana, Pierre-Loïc Garoche, Arnaud Dieumegard, Éric Féron, and Marc Pantel. "From design to implementation: An automated, credible autocoding chain for control systems." In: *Advances in Control System Technology for Aerospace Applications.* Ed. by Éric Féron. Vol. 460. Lecture Notes in Control and Information Sciences. Springer Berlin Heidelberg, 2016, pp. 137–180. ISBN: 978-3-662-47693-2. DOI: 10.1007/978-3-662-47694-9_5.

[148] Patrick Baudin, Jean-Christophe Filliâtre, Claude Marché, Benjamin
       Monate, Yannick Moy, and Virgile Prevosto. *ACSL: ANSI/ISO C Spec-
       ification Language*. `frama-c.cea.fr/acsl.html`. 2008.

[149] "Inquiry board traces Ariane 5 failure to overflow error." In: *SIAM News*
       29.8 (1996). available on internet archive, pp. 12–13.

Index

Acknowledgments

This work was made possible thanks to the generous support of French Agence Nationale de la Recherche and USA National Science Foundation through grants ANR-12-INSE-0007 (Combining Analyses for the Study of Numerical Invariants - CAFEIN), ANR-12-ASTR-0004 (Verification of Fast Optimization algorithms applied in Critical Embedded Control), ANR-17-CE25-0018 (Formal and Exhaustive Analyses of Numerical Intensive Control Software for Embedded Systems - FEANICSES), CNS1135955 (Credible Autocoding and Verification of Embedded Software - CrAVEs), and CNS1446758 (Semantics of Optimization for Real Time Intelligent Embedded Systems - SORTIES).

In addition, the long term support of both Onera, Information Processing and Systems Department and NASA, Aeronautics Research Mission Directorate, within the NASA Ames Robust Software Engineering group provided the perfect environment to develop that research.

Last, the presented work is only the result of fruitful and friendly collaborations. I would like then to express specifically my gratitude to: Behçet Açıkmeşe, Assalé Adjé, Maxime Arthaud, Hamza Bourbouh, Guillaume Brat, Adrien Champion, Alexandre Chapoutot, Raphaël Cohen, Sylvain Conchon, Guillaume Davy, Rémi Delmas, Eric Féron, Christophe Garion, Eric Goubault, Arie Gurfinkel, John Hauser, Didier Henrion, Heber Herencia-Zapana, Romain Jobredeaux, Temesghen Kahsai, Victor Magron, Matthieu Martel, Marc Pantel, Marc Pouzet, Sylvie Putot, Pierre Roux, Yassamine Seladji, Xavier Thirioux, Cesare Tinelli, Arnaud Venet, Timothy Wang, Alexis Werey, and Virginie Wiels.

PRINCETON SERIES IN APPLIED MATHEMATICS

Chaotic Transitions in Deterministic and Stochastic Dynamical Systems: Applications of Melnikov Processes in Engineering, Physics, and Neuroscience, Emil Simiu

Selfsimilar Processes, Paul Embrechts and Makoto Maejima

Self-Regularity: A New Paradigm for Primal-Dual Interior-Point Algorithms, Jiming Peng, Cornelis Roos, and Tamás Terlaky

Analytic Theory of Global Bifurcation: An Introduction, Boris Buffoni and John Toland

Entropy, Andreas Greven, Gerhard Keller, and Gerald Warnecke, editors

Auxiliary Signal Design for Failure Detection, Stephen L. Campbell and Ramine Nikoukhah

Thermodynamics: A Dynamical Systems Approach, Wassim M. Haddad, VijaySekhar Chellaboina, and Sergey G. Nersesov

Optimization: Insights and Applications, Jan Brinkhuis and Vladimir Tikhomirov

Max Plus at Work, Modeling and Analysis of Synchronized Systems: A Course on Max-Plus Algebra and Its Applications, Bernd Heidergott, Geert Jan Olsder, and Jacob van der Woude

Impulsive and Hybrid Dynamical Systems: Stability, Dissipativity, and Control, Wassim M. Haddad, VijaySekhar Chellaboina, and Sergey G. Nersesov

Positive Definite Matrices, Rajendra Bhatia

The Traveling Salesman Problem: A Computational Study, David L. Applegate, Robert E. Bixby, Vašek Chvátal, and William J. Cook

Genomic Signal Processing, Ilya Shmulevich and Edward R. Dougherty

Wave Scattering by Time-Dependent Perturbations: An Introduction, G. F. Roach

Algebraic Curves over a Finite Field, J.W.P. Hirschfeld, G. Korchmáros, and F. Torres

Distributed Control of Robotic Networks: A Mathematical Approach to Motion Coordination Algorithms, Francesco Bullo, Jorge Cortés, and Sonia Martínez

Robust Optimization, Aharon Ben-Tal, Laurent El Ghaoui, and Arkadi Nemirovski

Control Theoretic Splines: Optimal Control, Statistics, and Path Planning, Magnus Egerstedt and Clyde Martin

Matrices, Moments and Quadrature with Applications, Gene H. Golub and Gérard Meurant

Graph Theoretic Methods in Multiagent Networks, Mehran Mesbahi and Magnus Egerstedt

Totally Nonnegative Matrices, Shaun M. Fallat and Charles R. Johnson

Matrix Completions, Moments, and Sums of Hermitian Squares, Mihály Bakonyi and Hugo J. Woerdeman

Modern Anti-windup Synthesis: Control Augmentation for Actuator Saturation, Luca Zaccarian and Andrew W. Teel

Stability and Control of Large-Scale Dynamical Systems: A Vector Dissipative Systems Approach, Wassim M. Haddad and Sergey G. Nersesov

Mathematical Analysis of Deterministic and Stochastic Problems in Complex Media Electromagnetics, G. F. Roach, I. G. Stratis, and A. N. Yannacopoulos

Topics in Quaternion Linear Algebra, Leiba Rodman

Hidden Markov Processes: Theory and Applications to Biology, M. Vidyasagar

Mathematical Methods in Elasticity Imaging, Habib Ammari, Elie Bretin, Josselin Garnier, Hyeonbae Kang, Hyundae Lee, and Abdul Wahab

Rays, Waves, and Scattering: Topics in Classical Mathematical Physics, John A. Adam

Formal Verification of Control System Software, Pierre-Loïc Garoche